Springer-Verlag Berlin Heidelberg GmbH

Yuko Giga-Hama Hiromichi Kumagai (Eds.)

Foreign Gene Expression in Fission Yeast:

Schizosaccharomyces pombe

Springer

Yuko Giga-Hama, Ph.D.
Research Center, Asahi Glass Co., Ltd.
1150. Hazawa-cho, Kanagawa-ku-,
Yokohama-shi
Kanagawa, 221
Japan

Hiromichi Kumagai, Ph.D.
Research Center, Asahi Glass Co., Ltd.
1150. Hazawa-cho, Kanagawa-ku-,
Yokohama-shi
Kanagawa, 221
Japan

ISBN 978-3-662-03474-3

Library of Congress Cataloging-in-Publication Data
Foreign gene expression in fission yeast *Schizosaccharomyces pombe* / [edited by] Yuko Giga-Hama, Hiromichi Kumagai.
p.cm. - (Biotechnology intelligence unit)
Includes bibliographical references and index.
ISBN 978-3-662-03474-3 ISBN 978-3-662-03472-9 (eBook)
DOI 10.1007/978-3-662-03472-9
1. *Schizosaccharomyces pombe*. 2. Yeast fungi–Biotechnology. 3. Yeast fungi–Genetic engineering.
4. Recombinant proteins. I. Giga-Hama, Yuko, 1959-. II. Kumagai, Hiromichi, 1954. III. Series.
TN 248.27.Y43F67 1997
660'.65–dc21 97-19696 CIP

Originally published by Springer-Verlag Berlin Heidelberg New York in 1997
Softcover reprint of the hardcover 1st edition 1997

Typesetting: Landes Bioscience Georgetown, TX, U.S.A.

SPIN:10634479 31/3111 - 5 4 3 2 1 0 - Printed on acid-free paper

PREFACE

The present state of yeast expression technology holds much promise for the production of pharmacological target proteins and various industrial enzymes. Yeast expression systems have enabled the successful production of therapeutic proteins such as GM-CSF, hepatitis B virus, and human albumin, to name only a few. There is much commercial interest in yeast expression systems because of their advantages over mammalian cells and *E. coli*. The major focus of this book is to show the usefulness of the fission yeast, *Schizosaccharomyces pombe* (*S. pombe*) expression systems.

S. pombe cells are cylinders about 3-4 μm in diameter and 7-15 μm in length. They divide by medial fission, which shows that *S. pombe* is not closely related to budding yeast. The word "pombe" means beer in Swahili; the yeast was originally isolated from eastern African beer by P. Lindner in 1893.

S. pombe is the most intensely studied and well characterized of the yeast species aside from *Saccharomyces cerevisiae* (*S. cerevisiae*); they are harmless, rapidly growing eukaryotes that are frequently used as model systems to investigate various eukaryotic phenomena. The haploid *S. pombe* genome contains more than 7,000 genes in three chromosomes (14 million base pairs). Its complete DNA sequence is being determined by several laboratories.

Fission yeast is increasingly attractive as an experimental system for investigating various gene functions, especially regarding cell growth and division. *S. pombe* may be more suitable to complement mutant yeast homologues than is *S. cerevisiae* since: (1) comparisons of its DNA sequences have suggested that *S. pombe* may be more similar to mammalian cells than to *S. cerevisiae*, and (2) in phylogenetic analyses using nuclear small subunit rRNAs, *S. pombe* was found to branch before the divergence of budding yeasts such as *S. cerevisiae*, *Candida albicans* and *Pichia pastoris*, and filamentous ascomycetes including *Neurospora*, *Penicillium* and *Aspergillus*.

The utility of *S. pombe* in isolating mammalian genes by complementation of corresponding mutant homologues has been demonstrated, including Cdc2 kinase, Wee1 kinase and Cdc25 phosphatase. Recent advances in genetic and biochemical techniques will enable the further study of the mating, meiosis, mitosis and cell cycle of *S. pombe*.

S. pombe can be used as a model system for the expression of heterogeneous proteins from higher eukaryotes because it resembles higher eukaryotes in various aspects. Post-translational modification in *S. pombe* is also closer to that in higher eukaryote cells than to that in *S. cerevisiae*, i.e., *S. pombe* requires the lipid modification of proteins during the transport process in a cell and the presence of galactosyltransferase. However, the development of useful vectors for the fission yeast has lagged behind that for *S. cerevisiae* and *Pichia pastoris* for many years.

The characteristics of host-vector systems active in *S. pombe* are discussed in chapter 2 and the heterologous proteins successfully being expressed in this host are described.

Compared with the *E. coli* expression systems, one disadvantage of *S. pombe* expression systems has been the lower yield of product. The use of an efficient expression system such as pREP (chapter 3) and pTL2M1 vectors (chapter 4) has provided a solution to this problem.

For further application, it is important to discuss the considerations concerning industrial scale-up and fermentation optimization using various recombinant *S. pombe* to obtain high level expression. However, it is not clear whether *S. pombe* has the ability to reach very high cell densities compared with that of *Pichia pastoris*. The further development of *S. pombe* expression systems is needed to determine the optimum culture conditions for high-density cultures and for inducible promoter systems.

Aside from its commercial uses, *S. pombe* expression technology is also very important for genetic analyses. Its availability of genetic information will provide a firm basis for a better understanding of complex eukaryotic pathways including the cell cycle and other mechanisms.

The further development of *S. pombe* expression systems as a tool to produce useful proteins in large quantities is awaited for both academic and commercial use.

We express our gratitude to Professor Hiroto Okayama, the University of Tokyo, and all of the members of the Biochemical Group of the Asahi Glass Company.

General information about *S. pombe* including plasmids and expression vectors can be found at the Forsburg Lab Home Page

(http://www.flosun.edu/~forsburg/lab.html) and the University of Amsterdam home page (http://www2.bio.uva.nl/pombe/)

Yuko Giga-Hama
Hiromichi Kumagai
Research Center, Asahi Glass Co., Ltd.

CONTENTS

EDITORS

Yuko Giga-Hama, Ph.D.
Senior Staff Researcher
Research Center
Asahi Glass Co., Ltd.
Yokohama, Kanagawa, Japan
chapters 1 and 11

Hiromichi Kumagai, Ph.D.
Leader, Biotechnology Group
Research Center
Asahi Glass Co., Ltd.
Yokohama, Kanagawa, Japan
chapter 2

CONTRIBUTORS

Volker Hildebrandt, Ph.D.
UFZ-Umweltforschungszentrum Leibzig-Halle GmbH
Projektträger Wissenchaft-Technologie-Umwelt
Magdeburg, Germany
chapters 5 and 6

Kinsey Maundrell, Ph.D.
Head, Molecular Microbiology Section
Glaxo Wellcome Research and Development S.A.
Geneva, Switzerland
chapter 3

Dieter Oesterhelt, Ph.D.
Professor of Biochemistry
Max-Planck-Institut für Biochemie
Abt. Membranbiochemie
Martinsried, Germany
chapter 5

Paul Russell, Ph.D.
Associate Professor
Departments of Molecular Biology and Cell Biology
The Scripps Research Institute
La Jolla, California, U.S.A.
chapter 9

Kumar Sambamurti, Ph.D.
Associate Consultant
The Mayo Clinic
Jacksonville, Florida, U.S.A.
chapter 10

Toshiyuki Sasagawa, MD/Ph.D.
Department of Obstetrics and Gynecology
School of Medicine
Kanazawa University
Kanazawa, Ishikawa, Japan
chapter 8

Kazuhiro Shiozaki, Ph.D.
Research Associate
Department of Molecular Biology
The Scripps Research Institute
La Jolla, California, U.S.A.
chapter 9

Hideki Tohda
Research Scientist
Research Center
Asahi Glass Co., Ltd.
Yokohama, Kanagawa, Japan
chapter 4

Toshio Yasumori, Ph.D.
Assistant Professor
Department of Pharmacology
School of Medicine
Keio University
Tokyo, Japan
chapter 7

Part I

Characteristics of "Host" and "Vector"

CHAPTER 1

Fission Yeast *Schizosaccharomyces pombe*: An Attractive Host for Heterologous Protein Production

Yuko Giga-Hama

Introduction

The fission yeast *Schizosaccharomyces pombe (S. pombe)* is a unicellular eukaryote belonging to the Ascomycetes. *S. pombe* was originally isolated from East African millet beer. The millet beer, according to a secondhand report derived from the description by Saare in a German journal in 1890, consisted of "a light brown liquid, which had thick grayish-white settlings, covering half the bottle, with mixed-in pieces and remnants from the millet used. The taste was acidulous, similar to that of sour leaven from the distillery, rotten, and not particularly pleasant." Isolated *S. pombe* was sent from East Africa to Germany in 1890 and further cultivated to a pure culture. The first description of *S. pombe* by P. Lindner was published in a German brewery weekly journal[1] in 1893. He chose as its name the Swahili word for beer, "pombe." Microscopic examination revealed that the shape of each *S. pombe* cell is cylindrical, and rounded on the ends. Its length is 7-15 μm and its diameter is around 4 μm, yet it may fluctuate in size, depending especially on the culture conditions. *S. pombe* is called "fission yeast" because it only reproduces

Foreign Gene Expression in Fission Yeast Schizosaccharomyces pombe, edited by Yuko Giga-Hama and Hiromichi Kumagai.

by means of fission, besides by means of spores. No budding is observed such as occurs in brewer's yeast.

Unlike other yeast species, *S. pombe* has not been actively used to make products such as wine, beer and bread, although it was originally isolated from beer. However, among yeast species, *S. pombe*, along with a baker's yeast, *Saccharomyces cerevisiae* (*S. cerevisiae*), is most intensely studied and well characterized from molecular genetic and cell biological aspects. With the development of a transformation system of *S. pombe*,[2] many *S. pombe* genes have been identified and characterized, and this has led to the development of many molecular genetic manipulation methods for *S. pombe*.[3] With the progress in the *S. pombe* research on, this yeast has gradually come to be considered a useful species as an experimental model for elucidating many aspects of eukaryotic cells from the molecular biological standpoint. Indeed, *S. pombe* has been found to be one of the best experimental models for the study of cell cycle control. Other uses involve the study of chromosomal structure, signal transduction for sexual differentiation, and others.[3] These studies have emphasized how similar *S. pombe* and mammalian cells are and how distantly related this yeast and *S. cerevisiae* are.

Here we will compare *S. pombe* with other yeast species (mainly a baker's yeast, *S. cerevisiae*) and mammalian cells from several standpoints, and demonstrate that *S. pombe* has many attractive features for heterologous protein production.

The Life Cycle and Genome of *S. pombe*

Both *S. pombe* and *S. cerevisiae* belong to Ascomycetes, the members of which have special morphological features such as an ascus that contains ascospores produced by meiosis. Spores are formed in tetrads inside the cell walls. Both yeasts reproduce meiotically by ascospores. However, *S. pombe* proliferates asexually by cell division (fission), while *S. cerevisiae* reproduces mitotically by budding. The equatorial process of cell division characterized by septation and medial fission is more common among eukaryotes than in budding yeasts.

S. pombe and *S. cerevisiae* show both mitotic and meiotic cell cycles. Strictly speaking, however, there are dissimilarities between their cell cycles. *S. pombe* grows as a haploid with two mating types (h^+ and h^-) under normal conditions. Starvation induces mating, which results in a diploid zygote, followed by meiosis which forms four haploid ascospores that germinate under nutrient conditions.

In contrast, *S. cerevisiae* grows as a diploid. Starvation induces meiosis, which makes four haploid ascospores, followed by germination. Each haploid (α and a) type begins to mate under normal conditions. Pheromones are necessary for the haploids to mate with each other. The h^- type and h^+ type of *S. pombe* secrete mating pheromones called M-factor and P-factor, respectively,[4,5] while the *S. cerevisiae* pheromones are called α-factor and a-factor for the α type and a type haploids, respectively.[6] Similarly to *S. cerevisiae*, the homothallic h^{90} type of *S. pombe* in the natural world, each colony of which contains an equal ratio of h^+ and h^-[7], can switch its mating type. Both h^- and h^+ types were originally isolated for research and are called heterothallic.

Regarding the cell cycle and mitosis, *S. pombe* is more similar to higher eukaryotes than to *S. cerevisiae.*[8] During mitosis, the chromosomes of *S. pombe* condense, the mitotic spindle is present for a short time and the spindle pole body is undivided until the spindle appears.[9,10] These characteristics are typical of higher eukaryotes.[11,12] *S. cerevisiae* cells form a spindle early (in the late G1/S stage) and do not show significant chromosome condensation.[13] In the regulation of the cell cycle, *S. pombe* regulates the cycle at the G1 to S and G2 to M transitions, as do higher eukaryotes, while *S. cerevisiae* regulates it mainly at the G1 to S transition phase.[14]

S. pombe has three chromosomes of 5.7 megabase (Mb)(I), 4.6 Mb (II) and 3.5 Mb (III) which carry 7,000-10,000 genes. Over 500 of these genes have been identified.[15,16] The genome of *S. cerevisiae* is arranged in 16 chromosomes, but its genomic size is similar to that of *S. pombe.*[17] The complete sequence of the *S. cerevisiae* genome in 16 chromosomes was made public in 1996.[18] The *S. pombe* genome project is now in progress, and the entire DNA sequence will be elucidated in the near future. The following groups are taking part in this project:

Chromosome I, Bart Barrell at the Sanger Centre, Cambridge, UK; chromosome II, Dick McCombie of The Lita Annenberg Hazen Genome Center at the Cold Spring Harbor Laboratory and Mitsuhiro Yanagida of Kyoto University, Japan (from the centromere to the cdc2 locus); chromosome III, Mitsuoki Morimyo of the Genome Research Group at the National Institute of Radiological Sciences, Chiba, Japan.

The two yeasts differ markedly regarding the centromere structure. The centromeres of *S. pombe* are about 100-1,000 times larger than those of *S. cerevisiae* and contain repeated sequences, as do the centromere regions of higher eukaryotic chromosomes.[19] The *S.*

cerevisiae centromeres are short and are not associated with repeated sequences.[20]

S. pombe from the Phylogenetic Standpoint

In analyses of the phylogenetic relationship between *S. pombe* and *S. cerevisiae*, sequence comparisons of their 18S rRNA and 5S RNA have been effective.[21-23] The comparison of 18S rRNA within Ascomycetes has revealed that *S. pombe* diverged very early, and *S. cerevisiae* then appeared by the next divergence.[24] Although they share the feature of unicellularity, these two yeasts are phylogenetically distinct.

From the results of phylogenetic analysis of 5S rRNA, assuming that a branch point for humans and *S. cerevisiae* diverged 1,200 ± 75 million years ago, the divergence between *S. pombe* and humans is calculated to occur 1,000 million years ago. [25,26] However, Huysmans et al[27] shows that the divergence between *S. pombe* and *S. cerevisiae* is overestimated. There are several contradictions about the evolutionary relationship between these two yeasts. The question as to whether *S. pombe* diverged from ascomycetes early or appeared later from the lineage leading to *Metazoa* is not clear from the only analysis of the nucleotide sequences of 5S rRNA.

Mitochondrial genome analysis is important to clarify the phylogenetical position of some species. The entire mitochondrial DNA of *S. pombe* has been sequenced.[28] Mitochondrial DNA comprises around 6-14% of total DNA, depending on its growth stage.[29] The mitochondrial genome of *S. pombe* is the smallest in size among the fungi and is similar to mammalian mitochondrial genomes.[30,31,32] Although it belongs to fungi, *S. pombe* has many distinct features in its mitochondria far removed from other fungi.[33,34]

Comparisons of the protein sequences between homologous proteins in *S. pombe* and *S. cerevisiae* have revealed that amino acid type identities and positions are close to those found in comparisons of yeast and mammalian proteins. There is a large evolutionary gap between *S. pombe* and *S. cerevisiae*, almost as large as that separating them both from animals. Some *S. pombe* proteins and genes are slightly more similar to mammalian cells than is *S. cerevisiae*; e.g., Ras,[35] β-tubulin,[11] and ribosomal proteins.[36-38] Overall, *S. pombe* proteins seem to be related equidistantly to those of *S. cerevisiae* and those of mammalian cells.

These findings, derived from molecular techniques such as DNA, RNA and protein sequence analyses, demonstrate that *S. pombe* is phylogenetically as distant from budding yeast as it is from mammals. *S. pombe* probably diverged from the lineage of *S. cerevisiae* very early to form an ancestral branch of the Ascomycetes (Fig. 1.1).[39-41]

S. pombe Compared with Other Yeast Species and Mammalian Cells: Other Aspects

In addition to phylogenetic differences, other features of *S. pombe* which show greater similarity between mammalian cells and *S. pombe* than between mammalian cells and *S. cerevisiae* have been reported. The following sections describe features of *S. pombe* regarding the functional substitution of mammalian genes, DNA replication and transcription initiation mechanism, mRNA splicing, codon usage, a post-translational modification, protease and others.

Some Mammalian Genes Are Active in S. pombe

S. pombe is more tolerant to foreign genetic signals than is *S. cerevisiae*, enabling easier cloning of the genes of higher eukaryotes; this is probably because of the differences in gene expression and signal recognition mechanisms of both yeasts. Certain *S. cerevisiae* genes can be expressed functionally and complement defective alleles of their counterparts in *S. pombe*, but *S. pombe* genes are not expressed in *S. cerevisiae*.[42,43] Some mammalian genes can be isolated using *S. pombe* by complementation of the mutant homologue. The functional substitution of a human homologue of the cell

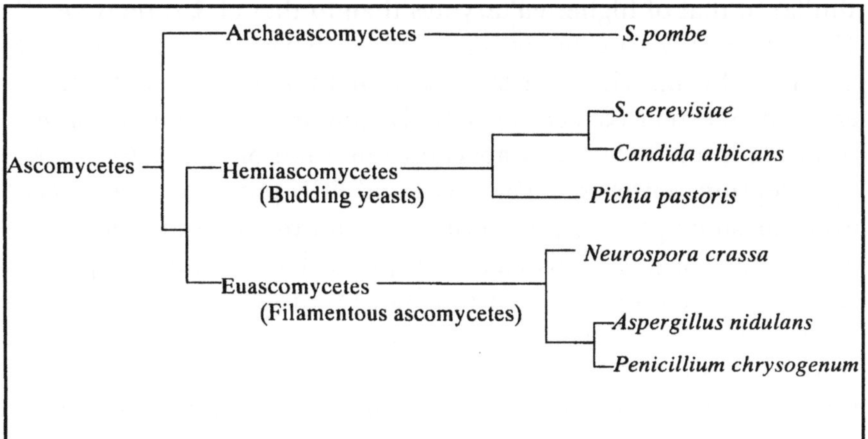

Figure 1.1 Phylogenetic tree of Ascomycetes[39-41]

Fig. 1.1.

cycle regulator protein cdc2 gene for the *S. pombe* cdc2 gene (which is homologous to the *S. cerevisiae* CDC28 gene) can be performed.[44] The similarity of the human cdc2 system to that of *S. pombe* has been confirmed at the protein level.[45]

For screening the human cDNA expression library, *S. pombe* may frequently be more suitable than *S. cerevisiae* because of its similarities to mammalian cells and the availability of expression vectors using mammalian promoters, as described below.

DNA Replication and Transcription Initiation Mechanism of S. pombe

As for the replication of *S. pombe* DNA, the specific DNA sequences termed "ARS" (autonomously replicating sequences) in the chromosomes of *S. pombe*, which can promote the autonomous replication and extra chromosomal maintenance of plasmids, have been isolated.[46-48] Although a sequence similar to an 11 bp ARS consensus sequence of *S. cerevisiae* exists in the *S. pombe* ARS fragments, the sequence is not essential for ARS activity.[46] The region required for ARS function consists of several hundred base pairs which are much longer than those (less than 100 bp) in *S. cerevisiae*.[46,47] The human replication origin is longer than 10 kb and no ARS consensus sequence has been identified in either *S. pombe* or human.[49] Both human and *S. pombe* likely have similar mechanisms for the start of DNA replication.[49] Allshire reported that *S. pombe* chromosome III can replicate and be maintained in mouse cells.[50]

The transcription initiation mechanism of *S. pombe* is more similar to that of higher eukaryotes than to that of *S. cerevisiae*.[51-53] The *S. pombe* TATA element is located 25-30 base pairs upstream of the start of transcription, exactly as in mammalian cells.[54] Promoters from *S. cerevisiae* genes generally function poorly in *S. pombe*, resulting in inefficient and aberrant initiation of transcription. As an exception, however, ADH1, CYC1 and PGK promoters have been used for some protein productions,[55,56] and the *S. cerevisiae* genes for three glycosidases have all been expressed in *S. pombe* using their respective promoters.[57] A mammalian promoter, SV40 early promoter, is functional in *S. pombe*. Transcripts initiate at the same position as in mammalian cells with this promoter.[58] Toyama et al[59-61] reported that some mammalian promoters were functional in *S. pombe*; the human chorionic gonadotropin α and human cytomegalovirus promoters are 10-fold stronger than SV40 early promoter.

With the use of a cytomegalovirus promoter, a high level expression vector has been constructed, and some proteins were successfully produced by this vector (chapter 4).

mRNA Maturation

S. pombe genes have intervening sequences, and several introns per gene are usually distributed throughout the coding region, whereas *S. cerevisiae* genes, in general, rarely possess introns.[62,63] *S. pombe* has several introns in its genes for β-tubulin,[11] ras homologue,[35] calmodulin,[64] and cell cycle regulator protein cdc2,[65] whereas *S. cerevisiae* has none.[66] In this respect, *S. pombe* resembles the higher eukaryotes more than does *S. cerevisiae*. However, the size of the *S. pombe* intron is smaller (40-120 bp)[63] than those of *S. cerevisiae* (250 to 550 bp)[62] and those of higher eukaryotes, which are variable in size but mostly longer than *S. pombe*.

As for the RNA splicing mechanism, *S. cerevisiae* s iather different from other fungi, including *S. pombe*, and from higher eukaryotes. The splicing system in *S. cerevisiae* is unique. It has severe consensus intron signal sequences such as the 5' intron/exon junction A(T)GGTAPyGTA(T) and the heptanucleotide TACTAAC,[67] which is located near the 3' end of the intron. These sequences are very conservative. In *S. pombe*, the intron signal sequences are not so stringent. CTNAC, located near the 3' end of the intron, is a degenerate version of the TACTAAC sequence in *S. cerevisiae* [65] and is similar to the consensus found in higher eukaryotes.[68] The intron signal sequences in *S. pombe* are variable, as in higher eukaryotes, but this variability is not the same as that found in higher eukaryotes[63,69] The introns in *S. pombe* are of a simple type compared with those of *S. cerevisiae* and higher eukaryotes.[70]

The components of the spliceosomes which remove the introns in *S. pombe* are rather similar to those of mammalian species.[67] For example, the small nuclear ribonucleoproteins (snRNP) U1 and U2 in *S. pombe* contain U1 and U2 snRNA, respectively, whose size and structure resemble those of human U1 and U2 snRNA rather than those of *S. cerevisiae*.[71,72] Regarding snRNA,[73] *S. pombe* seems to have retained a complement of snRNAs much more similar to mammalian types, both in number and sequence homology, while in *S. cerevisiae* they have diverged to the extent that none are present in the same amounts as in the mammalian sequences.

For these reasons, when higher eukaryotic genes containing introns are introduced into *S. cerevisiae*, they may not be expressed, while the same genes can be expressed when they are put in *S. pombe*. The splicing system in *S. pombe* removes one intron of the SV40 gene encoding the small T antigen, although the efficiency is low.[74]

The mechanism of mRNA 3' processing seems to be similar in yeast and in higher eukaryotes.[75] However, there are differences in the recognition signals in the pre-mRNA.[76] In higher eukaryotes, a highly conserved AAUAAA sequence is critical for polyadenylation, and the poly (A) tail is added to the 3' end of the precursor RNA generated by endonucleolytic cleavage. Both *S. pombe* and *S. cerevisiae* lack the AAUAAA motif and have no apparent consensus sequence. Additionally, the mammalian signal is not always functional.[76] The sequence surrounding the polyadenylation sites of yeasts are AT-rich, and this region of *S. cerevisiae* is functional in *S. pombe*.

Codon usage

Russell reported that degenerate codons are not used at equal frequencies in *S. pombe*.[3] Highly expressed genes in *S. pombe* such as *adh* (alcohol dehydrogenase) and *tpi* (triose-phosphate isomerase) have the most extreme codon usage bias. Russell proposed a Codon Bias Index (CBI) and showed that there is a strong correlation between high mRNA levels and a high CBI value.[3] Forsburg reported that the overall codon usages of *S. pombe* and *S. cerevisiae* are very similar to each other and differ from those of *Aspergillus* and humans.[77] Nakamura reported an up-to-date codon usage table for *S. pombe* and showed that the GC content of the coding sequence is 39.99%. (http://www.dna.affrc.go.jp/~nakamura/codon.htm1)

Sugar Chains in S. pombe

Cell Wall

Most yeasts have cell walls which are composed of a mixture of glucan and mannan macromolecules. The cell wall of *S. pombe* differs from that of *S. cerevisiae* as follows. The former contains alkali-soluble α-(1-3)-linked glucan, galactomannan instead of mannan, a large amount of O-linked sugar in the wall glycoprotein[78] and a very small amount of glucosamine. Glucosamine builds up chitin in the septa of *S. cerevisiae*,[79] whereas none or a small proportion (0.5%) of chitin is present in the septa of *S. pombe*, despite the presence of chitin synthase.[80,81] The cell walls of *S. pombe* are composed of galactomannan, 9-14%; alkali-soluble α-(1-3)-linked glucan, 18-28%; alkali-soluble β-(1-3)-linked glucan, 24%; β-(1-6)-linked glucan, 2%; and alkali-insoluble β-(1-3) -linked glucan.[82] The galactomannan is observed as an outer layer of the wall and adjacent to the plasma membrane.[83] The interior skeleton of the wall is organized as a network of two different kinds of glucan micro-fibrils, and amorphous glucan appears to occur both on the cell surface and within the wall.[84]

Glycoproteins

Like glycoproteins in animal cells, yeast glycoproteins contain carbohydrate chains of N-linked oligosaccharides using asparagine residues and of O-linked species using serine/threonine residues. In contrast to the glycoproteins derived from animal cells, yeast glycans contain outer chains consisting of mainly mannose oligomers in both N- and O-linked oligosaccharides. Unlike *S. cerevisiae* and other yeast species, *S. pombe* has galactose residues in mannose-rich sugar chains.

In mammalian cells, the processing of N-linked oligosaccharides occurs as follows. In the endoplasmic reticulum of mammalian cells, N-linked Glc3Man9GlcNAc2 is first trimmed to a Man8GlcNAc2 core structure and then processed by a series of mannosidases and sugar transferases to a final complex structure containing mannose, additional GlcNAc, galactose and sialic acid in the Golgi apparatus.[85]

In the N-linked glycan synthesis in *S. pombe* the smallest N-linked oligosaccharide core is Man9GlcNAc2, not Man8GlcNAc2, and is attached by 50-100 sugar residues consisting of both mannose and galactose (30% of the mass).[86] In *S. cerevisiae*, a common precursor is Man8GlcNAc2, elongated by the addition of more than 50 sugar residues consisting of only mannose.[87,88] In both yeasts, no additional GlcNAc or sialic acid is found. Galactose, which is not present in *S. cerevisiae*, is found in glycoproteins in *S. pombe*;[89,90] it is largely found at the non-reducing termini of the oligosaccharide chains.[86,91] *S. pombe* acid phosphatase, which is one of the cell surface-glycoproteins, contains not only mannose but also galactose in its carbohydrate moieties,[88] whereas only mannose is found in *S. cerevisiae* acid phosphatase. Purified *S. pombe* invertase was reported to contain carbohydrates by 67% of its total mass, composed of equal moles of mannose and galactose and a small amount of glucosamine.[89] Unlike *S. cerevisiae*, the detailed structure of the sugar chains of purified *S. pombe* glycoproteins has not yet been analyzed. *S. pombe* invertase, with a large amount of sugars, has recently been cloned and purified (Kaoru Takegawa, Kagawa University, Japan; personal communication). The structure of its oligosaccharide chains will be elucidated in the near future.

In higher eukaryotes, many kinds of glycoproteins are synthesized and secreted with various kinds of glycosyltransferases. These enzymes are membrane proteins, present especially in the endoplasmic reticulum and Golgi apparatus. As in higher eukaryotes, the modifications to the core structure occur in the Golgi apparatus in *S. pombe*.[92] *S. pombe* has a well-organized and morphologically distinct Golgi complex,[93] while *S. cerevisiae* has poor Golgi morphology. Chappell and Warren[94] demonstrated the presence of membrane-localized galactosyltransferase activity in *S. pombe* with substrate specificity capable of generating the α-linked galactose residues found on the N-linked oligosaccharides of invertase and acid phosphatase. Chappell et al[93] cloned a gene encoding an α1, 2 galactosyltransferase activity from *S. pombe* and showed that the galactosyltransferase activity is located in the Golgi apparatus in *S. pombe*. No galactosyltransferase has been found in *S. cerevisiae*. A deletion strain of the galactosyltransferase was found to significantly reduce the level of galactosyltransferase activity.[92] The oligosaccharide processing in *S. pombe* is more complex and the morphology and function of the *S. pombe* Golgi apparatus are more similar to mammalian cells than those of *S. cerevisiae*.[94]

Other Post-Translational Modifications

In mammalian cells, the amino-terminal modification of proteins in the cytosol is the most common processing event. The N-acetylation of proteins such as melanocyte-stimulating hormone and β-endorphin is required for full biological activity.[95] Met amino-peptidase (MAP) removes the N-terminal Met residue, and N-acetyltransferase (NAT) catalyzes the acetylation of the N-terminal residue, both of which are associated with ribosomes. The specificity of these enzymes appears to be conserved among eukaryotes,[96,97] and the protein acetylated in mammalian cells might be expected to be acetylated in yeast cells. Indeed, human lipocortin I, which is a N-acetylated protein, was expressed in *S. pombe* with its acetylated N-terminus (chapter 4).

The isoprenylation of proteins including G proteins and ras proteins, which plays an important role in the intracellular targeting of proteins as well as their functional activation, has recently been observed.[98] Modifications occur at the carboxy-terminal motifs of the general type Cys-A-A-X-COOH (A, aliphatic; X, any other amino acid). This cysteine residue is modified by a polyisoprenoid such as farnesyl (15 carbon) and geranylgeranyl (20 carbon), followed by the removal of three C-terminal residues. Cys-COOH is then methyl-esterified. Polyisoprenoids are derived from mevalonic acid. All mammalian cells can take up exogenous mevalonic acid with varying efficiencies. *S. pombe* has a high capacity for the uptake of mevalonic acid, and both geranylgeraniol and farnesol exist in total protein-bound isoprenoids in similar amounts.[99] One of the *S. pombe* mating factors, M-factor[5], is a nonapeptide the carboxy-terminal cysteine residue of which is both S-farnesylated and carboxy-methylated. *S. cerevisiae*, in contrast, is not suitable for studies on polyisoprenylation of proteins because of its inability to mediate the incorporation of mevalonic acid into polyisoprenoid moieties of proteins, due to the poor uptake of exogenous mevalonic acid.[98]

The N-terminal Gly of proteins such as G proteins and src tyrosine kinases are myristylated, and the proteins such as *fos* and *c-myc* are phosphorylated in mammalian cells. Such myristylation and phosphorylation are reported to occur in both *S. cerevisiae*[100-102] and *S. pombe.*[103,104]

Proteases

Not many proteases have been detected and isolated from *S. pombe* compared to *S. cerevisiae*, in which many processing proteases and vacuole proteases, among others, have been detected.[105] *S. pombe* probably has proteases like those in *S. cerevisiae.* Suaréz-Rendueles reported that two endoproteinases, one carboxypeptidase, one aminopeptidase and one dipeptidyl-aminopeptidase were found in *S. pombe*; intracellulary protease inhibitors were also detected.[106] Several of the isolated proteases are described below.

Carboxypeptidase Y in the vacuole was cloned.[107] It has amino acids 45% identical to those of *S. cerevisiae*. A serine protease homologous to *S. cerevisiae* protease B in the vacuole was isolated from *S. pombe.*[108,109] The morphology of the vacuole of *S. pombe* differs from that of *S. cerevisiae* and is similar to the lysosome of mammalian cells.[110] A deficient mutant of Carboxypeptidase Y or protease B homologue was isolated.[109,110] A dibasic endopeptidase in the Golgi aparatus was isolated—Krp1, which is homologous to Kex2, one of *S. cerevisiae* processing enzymes.[111] Sxa1 and Sxa2, which are putative proteases involved in pheromone communication, have been identified.[112,113] Arbesu purified YspI, a dipeptidyl aminopeptidase;[114] it is mapped on chromosome III.[115]

In cases of foreign gene expression, proteolytic activity is critical. The correct processing of the signal peptide or prepro region is necessary for the secretion of mature proteins. However, undesirable processing events may occur as a result of cleavage by a protease. In some cases, protease-deficient yeasts are advantageous for protein production.

Others

Signal Transduction

S. pombe possesses a signal transduction system with marked similarities to mammalian G -protein-coupled systems. Xu[116] reported that signals initiated by the mating factors of *S. pombe* are also transmitted through a G protein to the effector(s). Moreover, Obara[117] found that *S. pombe* utilizes a G protein subunit to positively transmit the signal from the mating factor receptor to the effectors. This resembles the role of Gα during signal transduction in mammalian cells, and makes *S. pombe* an appropriate model system for investigating mammalian receptor function.

Heat Shock Factor

The heat shock response of eukaryotic cells appears to have been conserved throughout evolution. In *S. cerevisiae*, a heat shock factor (HSF) is constitutively bound to the heat shock element (HSE), and heat shock increases the degree of transcriptional stimulation attributed to the HSF. In contrast, the *S. pombe* HSF is fully induced only under stressed conditions in a manner similar to that of *Drosophila* and mammalian cells, but different from the constitutive binding by HSF in *S. cerevisiae.*[118,119] The ability of *S. pombe* HSF to bind to the HSE significantly increases upon heat shock. The structure of the *S. pombe* HSF gene is similar to that of *Drosophila* and mammalian HSF. The similarities between the regulation of HSF in *S. pombe* and mammalian cells further contribute to the usefulness of *S. pombe*, as a model for the study of heat-induced HSF.

The comparisons of *S. pombe* and *S. cerevisiae* from the various standpoints described above show that although similar manipulation techniques can be applied to both yeasts, *S. pombe* is not closely related to *S. cerevisiae* in many respects. *S. pombe* has many features in common with higher eukaryotes, making fission yeast an informative and accurate eukaryotic model system for molecular biology studies. There are also prospective biotechnological applications of this microorganism, such as heterologous eukaryotic gene expression.

Development of foreign gene expression systems in *S. pombe*

Background

The production of heterologous proteins utilizing recombinant DNA technology has been conducted in prokaryotic and eukaryotic microorganisms or cells such as *Escherichia coli (E. coli)*, *Bacillus*, *Streptomyces*, *Aspergillus*, yeast, animals (including insects) and plant cells.[120] As foreign proteins, various biogenic polypeptides are considered to be accessible, and many of them have been produced on a largescale for medical applications in addition to their use in research.

However, methods employing prokaryotes are not effective for all polypeptides, and it is not always easy to reproduce the complicated post-translational modification of eukaryotic proteins and to reproduce the natural steric structures. Actually, some proteins produced by prokaryotic cells are inapplicable for medicines due to their irregular structures and activities.[121] In addition, *E. coli* has toxic cell wall pyrogens, a characteristic endotoxin, which might contaminate end products.

In animal or plant cells, protein production efficiency is low, because these cells are more difficult to handle than microorganisms, their culture is costly, and they are obtainable only at low cell densities. In some cases mammalian cells may contain oncogenic or viral DNA, so that products from these organisms must be tested more extensively.

For these reasons, yeasts are the ideal organisms for the production of foreign proteins, especially eukaryotic polypeptides. Yeasts are advantageous in light of their established culture methods and their ability to express the genetic information of eukaryotes. They can grow rapidly to high cell density in simple media. Their genetics are more advanced than those of any other eukaryotes, and they can be manipulated easily, as is the case for *E. coli*. Yeasts have been conventionally used in the fermentation and food industries, and have proven to be safe for human consumption, since they have no endotoxins. They are quite appropriate for the production of proteins for medical use.

The Status Quo of Yeast Expression Systems

On the bases of this background, heterologous protein expression systems using yeasts continue to develop rapidly.[105] Transformation protocols have been established for many yeast species, and a number of alternatives to *S. cerevisiae* (which has often been used as a conventional host for protein expression for research, industrial or medical uses) have come to be used as host cells for foreign protein expression. The following yeasts are typical alternatives to *S. cerevisiae*:[105] *Pichia pastoris* [122] and *Hansenula polymorpha*,[123] methylotroph yeasts which were originally developed for the large-scale, high-yield production of heterologous protein in a medium containing methanol; *Kluyveromyces lactis*,[124] which has been used for many years to produce β-galactosidase (lactase); *Yarrowia lipolytica*,[125] used for many industrial processes such as the production of citric acid and isopropyl malate; and *Candida utilis*,[126] which is used to produce a number of commercially useful materials, such as glutathione, ribonucleic acids, amino acids and enzymes.

S. pombe has many features similar to those of higher eukaryotic cells, as described in this chapter, whereas the yeasts mentioned above are phylogenetically closer to *S. cerevisiae* than to *S. pombe*[24,40] The use of *S. pombe* as a host is therefore expected to provide foreign gene products closer to their natural form, like those produced by animal cells. Moreover, since the culture methods of different yeasts have much in common, knowledge about some yeasts can be easily applied to other yeasts.

Consequently, since *S. pombe* has many attractive characteristics and can be simply manipulated like other types of yeast, various kinds of proteins are expected to be produced by the expression system using *S. pombe* as a host, in addition to previously developed host-vector systems.

Foreign Gene Expression in S. pombe

Although *S. pombe* is a promising host for foreign gene expression, its biotechnological applications have not all been successful, and its use is far behind that of *E. coli* and *S. cerevisiae* in studies of recombinant DNA technology. *S. pombe* has a naturally occurring plasmid of rDNA and a 2 μm plasmid which is not homologous with the 2 μm plasmid of *S. cerevisiae*.[127] In *S. cerevisiae*, the *E. coli*-yeast shuttle vectors based on its 2 μm plasmid are the most commonly used vectors for foreign gene expression,[128] and many useful expression vectors have been developed so far.[105] An expression vector based on the *S. pombe* 2 μm plasmid has not yet been developed. Following the development of a transformation system,[2] a number of selective marker genes[129] and an ars-stb system[130] were identified in *S. pombe*, and several expression vectors have been constructed (chapter 2). Of these expression vectors, the inducible vectors developed by Maundrell (chapter 3) are most commonly used; the expression can be regulated by thiamine and the production level is high. Tohda et al recently developed (chapter 4) a constitutive high level expression vector with a powerful promoter. Its copy number is controlled by the concentration of antibiotics in the medium. Giga-Hama et al also developed a secretion vector for the foreign secretory protein in *S. pombe*, which has a signal sequence derived from the *S. pombe* mating pheromone, P-factor (chapter 11). Some proteins have thus been successfully produced, as described in Part III.

***Table 1.1. Production of heterologus proteins in* S. pombe**

Location	Protein	Reference
Cytosol	Capsid proteins of human papillomaviris	chap 8
	Cell-cycle regulatory proteins	chap 9; 131
	Human lipocortin I	chap 4
	Rat and human arginase	chap 4
	Rat NDP-kinase	chap 4
	Nematode galectin	chap4
	Human blood coagulation protein Factor XIIIa	132
	Polyoma middle T-antigen	133
	β-glucuronidase	134
	Green flourescent protein	135
	HIV-Type I protein R	136
Membrane	Human β2-adrenergic receptor	137
	human D2s dopamine receptor	138
	Neurokinin NK2 receptor and G protein components	139
	Human P-glycoprotein	140
	Human liver epoxide hydrolase	141
	Bacteriorhodopsin	chaps 5 and 6
	Chloroplast triose phosphate translocator	142
	Chlorella hexose transporter	143
	Cytochrome P_{450}	chap 7; 144
Extracellular	Human gastric lipase	145
	Mouse α-amylase	146
	Human antithrombin III	56
	Human interleukin-6	chap 11
	S.cervisiae invertase	147, 148
	Human placental alkaline phoshatase	chap 10

Heterologous Protein Production

Table 1.1 lists reported studies of heterologous protein production in *S. pombe*. Many proteins will be increasingly produced in *S. pombe*, together with the development of expression vectors.

The Properties of Produced Proteins in *S. pombe*

The examples described above show that *S. pombe* is an attractive candidate for producing not only cytosolic proteins but also mammalian membrane proteins and secreted proteins with native properties. Sander[138] first reported the expression of a G protein-coupled receptor in *S. pombe*, and showed that *S. pombe* is an even more suitable organism for the expression of the human D2s dopamine receptor. These membrane proteins were found to be inserted into the cell membrane. In some cases, the recombinant receptor production was at least 3 times higher in *S. pombe* than in *S. cerevisiae*. *S. pombe* has a signal transducing mechanism similar to those of mammalian G protein-coupled systems ("Others", above). Sande and Privalsky reported that the *S. pombe* system exhibits significant advantages over the *S. cerevisiae* system for the reconstitution of some hormone receptors.[149]

S. pombe is also considered to be an attractive host for secretory proteins. Glycosylation occurs in the lumen of the *S. pombe* endoplasmic reticulum (ER), as in mammalian cells. Glycoproteins are then transported into the Golgi apparatus where modifications take place, although these modifications differ from those which occur in higher eukaryotic cells ("Sugar Chains in *S. pombe*", above). After the modifications, the proteins are packaged into secretory vesicles and delivered to the cell surface. For the secretion of foreign proteins, the secretory signal is important. Some proteins can be secreted using a homologous signal sequence, while others are secreted using heterologous signals; for example, α-amylase was secreted using a signal sequence derived from *Kluyveromyces lactis* killer toxin,[146] and human interleukin (IL)-6 was secreted using the P-factor signal sequence (chapter 11). It is detailed in chapter 11 that the secretion signal derived from P-factor was accurately removed by processing enzymes, followed by efficient secretion. The *S. pombe* protease Krp1 ("Proteases", above), which is homologous to the protease Kex2 in the Golgi apparatus of *S. cerevisiae*, could play an important role in the processing of secretory proteins. The Golgi apparatus of *S. pombe* is morphologically similar to that of cultured mammalian cells, while *S. cerevisiae* possesses poor Golgi morphology ("Sugar Chains in *S. pombe*", above). These factors could make *S. pombe* superior for foreign secretory protein production, since some post-translational modifications occur in the Golgi apparatus.

Additionally, several analyses of the carbohydrate compositions of the heterologous glycoproteins expressed in *S. pombe* [56,146,147] and of the carbohydrate moieties of endogenous glycoproteins in *S. pombe* have revealed differences from *S. cerevisiae* ("Sugar Chains in *S. pombe*", above). These differences may also contribute to the usefulness of *S. pombe* for foreign glycoprotein expression.

Future Perspectives

Many more heterologous proteins are going to be produced in *S. pombe* with the use of the many types of expression vectors which have been and will be developed. The analysis of the produced proteins will clarify the detailed structure of the sugar chains of glycoproteins and the state of the other post-translational modifications, clarifying the mechanism of protein synthesis in *S. pombe* and the nature of *S. pombe* itself, and providing new information for the phylogeny and taxonomy of yeast species.

It has now become possible to produce foreign proteins, especially those derived from higher eukaryotes, at a high level by using *S. pombe* expression systems. High-level production, easy manipulation, and low cost, among other factors, make these expression systems attractive not only for academic research but also for industrial uses such as the production of medicines and enzymes, for example.

From this point onward many genes will be identified by research within the Human Genome Project, and the demand for protein production systems will be further increased in both academic and industrial fields. Many types of host-vector systems are needed, and protein production systems using *S. pombe* as a host are expected to be successfully utilized for many applications.

References

1. Lindner P. *Schizosaccharomyces pombe* n. sp., ein neuer Gährungserreger. Wochenschrift für Brauerei 1893; 10:1298-1300.
2. Beach D, Nurse P. High frequency transformation of the fission yeast *Schizosaccharomyces pombe*. Nature 1981; 290:140-142.
3. Russell P. Gene cloning and expression in fission yeast. In: Nasim A, Young P, Johnson BF, eds. Molecular Biology of the Fission Yeast. San Diego: Academic Press, Inc 1989:243-271.

4. Imai Y, Yamamoto M. The fission yeast mating pheromone P-factor: its molecular structure, gene structure, and ability to induce gene expression and G1 arrest in the mating partner. Gene & Dev 1994; 8:328-338.
5. Davey J. Isolation and quantitation of M-factor, a diffusible mating factor from fission yeast *Schizosaccharomyces pombe*. Yeast 1991; 7:357-366.
6. Thorner J. Pheromonal regulation of development in *Saccharomyces cerevisiae*. In: Strathern JN, Jones EW, Broach JR, eds. Molecular Biology of the Yeast *Saccharomyces:* Life Cycle and Inheritance. Cold Spring Harbor, New York: Cold Spring Harbor Laboratory. 1981:143-180.
7. Beach DH. Cell type switching by DNA transcription in fission yeast. Nature 1983; 305:682-687.
8. Nurse P. Cell cycle control genes in yeast. Trends Genet 1985; 1:51-55.
9. Nasmyth K, Nurse P. Cell division cycle mutants altered in DNA replication and mitosis in the fission yeast *Schizosaccharomyces pombe*. Mol Gen Genet 1981; 182:119-124.
10. Hagan IM, Hyams JS. The use of cell division cycle mutants to investigate the control of microtuble distribution in the fission yeast *Schizosaccharomyces pombe*. J Cell Sci 1988; 89:343-357.
11. Hiraoka Y, Toda T, Yanagida M. The NDA3 gene of fission yeast encodes β-tubulin: A cold-sensitive nda3 mutation reversibly blocks spindle formation and chromosome movement in mitosis. Cell 1984; 39:349-358.
12. McCully EK, Robinow CF. Mitosis in the fission yeast *Schizosaccharomyces pombe*: A comparative study with light and electron microscopy. J Cell Sci 1971; 9:475-507.
13. Erard M, Barker D, Geen J. Is chromosome condensation a phylogenetic marker? Stud Mycol 1987; 30:267-277.
14. Forsburg SL, Nurse P. Cell cycle regulation in the yeasts *S. cerevisiae* and *S. pombe*. Annu Rev Cell Biol 1991; 7:227-256.
15. Fan JB, Chikashige Y, Smith CL et al. Construction of a *Not*I restriction enzyme map of the fission yeast *Schizosaccharomyces pombe* genome. Nucl Acid Res 1989; 17:2801-2818.
16. Lennon GG, Lehrach H. Gene data base for the fission yeast *Scizosaccharomyces pombe*. Curr Genet 1992; 21:1-11.
17. Williamson DH. Nucleus; chromosomes and plasmids. In: Rose AH, Harrison JT, eds. The Yeasts. London: Academic Press 1991; 4:433-488.
18. Goffeau A, Barrell BG, Bussey H et al. Life with 6000 genes. Science 1996; 274:546-567.
19. Kuhn RM, Clarke L, Carbon J. Clustered tRNA genes in *Schizosaccharomyces pombe* centromeric DNA sequence repeats. Proc Natl Acad Sci USA 1991; 88:1306-1310.
20. Hegemann JH, Fleig U. The centromere of budding yeast. BioEssays 1993; 15:451-460.

21. Taylar JW, Bowman BH, Berbee ML et al. Fungal model organisms: Phylogenetics of *Saccharomyces, Aspergillus,* and *Neurospora.* Syst Biol 1993; 42:440-457.
22. Sipiczki M. Taxonomy and phylogenesis. In: Nasim A, Young P, Johnson BF, eds. Molecular Biology of the Fission Yeast. San Diego: Academic Press, Inc. 1989:431-452.
23. Spiczki M. Phylogenesis of fission yeasts. Contradictions surrounding the origin of a century old genus. Antonie van Leeuwenhoek 1995; 68:119-149.
24. Kurtzman CP, Robnett CJ. Phylogenetic relationships among species of *Saccharomyces, Schizosaccharomyces, Debaryoyomyces* and *Schwanniomyces* determined from partial ribosomal RNA sequences. Yeast 1991; 7:61-72.
25. Kimura H, Ohta T. Eukaryotes-prokaryotes divergence estimated by 5S ribosomal RNA sequences. Nature (London) New Biol 1973; 243:199-200.
26. Hori H, Osawa S. Evolutionary change in 5S RNA secondary structure and a phylogenetic tree of 54 5S RNA species. Proc Natl Acad Sci USA 1979; 76:381-385.
27. Huysmans E, Dams E, Vanderberghe A et al. The nucleotide sequences of the 5S rRNAs of four mushrooms and their use in studying the phylogenetic position of *Basidiomycetes* among the eukaryotes. Nucleic Acids Res 1983; 11:2871-2880.
28. Lang BF. The mitochondrial genome of *Schizosaccharomyces pombe.* In: O'Brien SJ, ed. Genetic Maps: Locus Maps of Complex Genomes. 6th ed, No. 3 Lower eukaryotes. New York: Cold Spring Harbor Laboratory Press. Plainview. 1993:118-119.
29. Bostock CJ. Mitochondrial DNA in the fission yeast *Scihzosaccharomyces pombe.* Biochim Biophs Acta 1969; 195:579-581.
30. Clark-Walker GD, Sriprakash KS. Sequence rearrangements between mitochondrial DNAs of *Torulopsis glabrata* and *Kloeckera africana* identified by hybridization with six polypeptide encoding regions from *Saccharomyces cerevisiae* mitochondrial DNA. J Mol Biol 1981; 151:367-387.
31. Anderson S, Bankier AT, Barrell BG et al. Comparison of the human and bovine mitochondrial genomes. In: Slonimski PP, Borst P, Attardi G, eds. Mitochondrial Genes. Cold Spring Harbor Laboratory, Cold Spring Harbor: New York. 1982:5-43.
32. Sankoff D, Leduc G, Antoine N et al. Gene order comparisons for phylogenetic inference; Evolution of the mitochondrial genome. Proc Natl Acad Sci 1992; 89:6575-6579.
33. Lang BF, Ahne F, Bonen L. The mitochondrial genome of the fission yeast *Schizosaccharomyces pombe.* The cytochrome b gene has an intron closely related to the first two introns in the *Saccharomyces cerevisiae cox1* gene. J Mol Biol 1985; 184:353-366.
34. Lang BF, Ahne F, Distler S et al. Sequence of the mitochondrial DNA, arrangement of genes and processing of their transcription in

Schizosaccharomyces pombe. In: Schweyen RJ, Wolf K, Kandewitz F, eds. Mitochondria. De Gruyer, Berlin. 1983:313-329.
35. Fukui Y, Kaziro Y. Molecular cloning and sequence analysis of a *ras* gene from *Schizosaccharomyces pombe*. EMBO J 1985; 4:687-691.
36. Otaka E, Higo K, Itho T. Yeast ribosomal proteins: VII, Cytoplasmic ribosomal proteins from *Schizosaccharomyces pombe*. Mol Gen Genet 1983; 191:519-524.
37. Otaka E, Ooi T, Itho T et al. Examination of protein sequence homologies: II Ribosomal protein YS25 from *Saccharomyces cerevisiae* and its counterparts from *Schizosaccharomyces pombe*, rat liver and *Escherichia coli*. J Mol Evol 1986; 23:337-343.
38. Moreno S, Klar A, Nurse P. Molecular genetic analysis of fission yeast *Schizosaccharomyces pombe*. Guthrie C, Fink GR eds. Methods in Enzyme. San Diego: Academic Press, Inc. 1991; 194:795-823.
39. Nishida H, Sugiyama J. Archiascomycetes. Detection of a major new lineage within the Ascomycota. Mycoscience 1994; 35:361-366.
40. Bruns TD, Vilgalys R, Barns SM, et al. Evolutionary relationships within the fungi: analysis of nuclear small subunit rRNA sequences. Mol Phylog Evol 1992; 1:231-241.
41. Berbee ML, Taylor JW. Dating the evolutionary radiations of the true fungi. Can J Bot 1993; 71:1114-1127.
42. Beach D, Durkacz B, Nurse P. Functional homologues cell cycle control genes in budding and fission yeasts. Nature 1982; 300: 706-709.
43. Johnston LH, Baker DG, Nurse P. Cloning and characterization of the *Schizosaccharomyces pombe* DNA ligase CDC17. Gene 1986; 41:321-325.
44. Lee MG, Nurse P. Complementation used to clone a human homologue of the fission yeast cell cycle control gene cdc2. Nature 1987; 327:31-35.
45. Draetta G, Brizuela L, Potashkin J et al. Identification of p34 and p13 human homologs o the cell cycle regulators of fission yest encoded by *cdc*$^{+}$ and *suci*$^{+}$ Cell 1987; 50:319-325.
46. Maundrell K, Huchinson A, Shall S. Sequence analysis of *ars* elements in fission yeast. EMBO J 1988; 7:2203-2209.
47. Dubey DD, Zhu J, Carlson DL et al. Three ARS elements contribute to the *ura4* replication origin region in the fission yeast, *Schizosaccharomyces pombe* EMBO J 1994; 13:3638-3647.
48. Okuno Y, Okazaki T, Masukata H. Identification of a predominant replication origin in fission yeast. Nuc Acid Res 1997; 25:530-536.
49. Masukata H. Replication origins in human and *S. pombe*. Cell Technology (Saiboukougaku in Japanese). 1996; 15:41-48.
50. Allshire RC, Cranston G, Gosden JR et al. A fission yeast chromosome can replicate autonomously in mouse cells. Cell 1987; 50: 391-403.
51. Russell P. Evolutionary divergence of the mRNA transcription initiation mechanism in yeast. Nature 1983; 301:167-169.

52. Russell P. Transcription of the triosephosphate isomerase gene of *S. pombe* initiates from a start point different from that in *S. cerevisiae*. Gene 1985; 40:125-130.
53. Fischli A, Schmid S, Coppolecchia R et al. The transcriptional initiation factor eIF4A from *Schizosaccharomyces pombe* is closely related to its mammalian counterpart. Yeast 1996; 12:977-981.
54. Benoist C, O'Hare K, Breathnach R et al. The ovalbumin gene-sequence of putative control regions. Nucleic Acid Res 1980; 8:127-142.
55. Belsham GJ, Barker DG, Smith AE. Expression of polyoma virus middle-T antigen in *Saccharomyces cerevisiae*. Eur J Biochem 1986; 156:413-421.
56. Bröker M, Ragg H, Karges HE. Expression of human antithrombin III in *Saccharomyces cerevisiae* and *Schizosaccharomyces pombe*. Biochem Biophys Acta 1987; 908:203-213.
57. Kuranda MJ, Robbins PW. Cloning and heterologous expression of glycosidase genes from *Saccharomyces cerevisiae*. Proc Natl Acad Sci USA 1987; 84:2585-2589.
58. Jones R, Moreno S, Nurse P et al. Expression of the SV40 promoter in fission yeast: identification and characterization of an AP1-like factor. Cell 1988; 53:659-667.
59. Toyama R, Okayama H. Human chorionic gonadotropin α and human cytomegalovirus promoters are extremely active in the fission yeast *S. pombe*. FEBS Lett 1990; 268:217-221.
60. Toyama R, Bende SM, Dhar R. Transcriptional activity of the human immunodeficiency virus-1 LTR promoter in fission yeast *Schizosaccharomyces pombe*. Nucleic Acid Res 1992; 20:2591-2596.
61. Swaminathan S, Malhotra P, Manohar CF et al. Activation of a dual adenovirus promoter containing nonconsensus TATA motifs in *S. pombe*:role for TATA sequence in the efficiency of transcription. Nuc Acid Res 1993; 11:2737-2746.
62. Woolford JL. Yeast pre-mRNA splicing. Yeast 1989; 5:439-457.
63. Prabhala G, Rosenberg GH, Kaufer NH. Architectural features of pre-mRNA intron in the fission yeast *Schizosaccharomyces pombe*. Yeast 1992; 8:171-182.
64. Takeda T, Yamamoto M. Analysis and *in vivo* disruption of the gene coding for calmodulin in *Schizosaccharomyces pombe*. Proc Natl Acad Sci USA 1987; 84:3580-3584.
65. Hindley J, Phear GA. Sequence of the cell division gene CDC2 from *Schizosaccharomyces pombe*; pattern of splicing and homology to protein kinases. Gene 1984; 31:129-134.
66. Davis TN, Urdea MS, Masiarz FR et al. Isolation of the yeast calmodulin gene: Calmodulin is an essential protein. Cell 1986; 47:423-431.
67. Guthrie L, Riedel N, Parker R et al. Genetic analysis of snRNAs and RNA processing in yeast. In: Hicks J, ed. Yeast Cell Biology Liss, New York 1986; 301-321.

68. Mount SM. A catalogue of splice junction sequences. Nucleic Acid Res 1982; 10:459-472.
69. Kishida M, Nagai T, Nakaseko Y et al. Meiosis-dependent mRNA splicing of the fission yeast *Schizosaccharomyces pombe mes1⁺* gene. Curr Genet 1994; 25:497-503.
70. Gatermann KB, Hoffmann A, Rosenberg GH et al. Introduction of functional artificial introns into the naturally intronless *ura4* gene of *Schizosaccharomyces pombe*. Mol Cell Biol 1989; 9:1526-1535.
71. Brennwald P, Porter G, Wise JA. U2 small nuclear RNA is remarkably conserved between *Schizosaccharomyces pombe* and mammals. Mol Cell Biol 1988; 8:5575-5580.
72. Porter G, Brennwald P, Wise JA. U1 small nuclear RNA from *Schizosaccharomyces pombe* has unique and conserved features and is encoded by an essential single-copy gene. Mol Cell Biol 1990; 10:2874-2881.
73. Hughes JM, Koninngs DAM, Cesaraeni G. The yeast homologue of U3 snRNA. EMBO J 1987; 6:2145-2155.
74. Käufer NF, Simanis V, Nurse P. Fission yeast *Schizosaccharomyces pombe* correctly excises a mammalian RNA transcript intervening sequence. Nature 1985; 318:78-80.
75. Wickens M. How the messenger got its tail: addition of poly (A) in the nucleus. Trends Biochem Sci 1990; 15:277-281.
76. Humphrey T, Sadhale P, Platt T et al. Homologous mRNA 3' end formation in fission and budding yeast. EMBO J 1991; 10:3503-3511.
77. Forsburg SL. Codon usage table for *Schizosaccharomyces pombe*. Yeast 1994; 10:1045-1047.
78. Font de Mora J, Valentin E, Herrero E et al. Glycoprotein molecules in the wall of *Schizosaccharomyces pombe* wild-type cells and a morphologically altered mutant resistant to papulacandin B. J Gen Microbiol 1990; 136:2251-2259.
79. Molano J, Bowers B, Cabib E. Distribution of chitin in the yeast cell wall. An ultrastructural and chemical study. J Cell Biol 1980; 261:15147-15152.
80. Horisberger M, Vonlanthen M, Rosset J. Localization of α-galactomannan and a wheat germ agglutinin receptors in *Schizosaccharomyces pombe*. Arch Microbiol 1978; 119:107-111.
81. Sietsma JH, Wessels JGH. The occurrence of glucosaminoglycan in the wall of *Scizosaccharomyces pombe*. J Gen Micrbiol 1990; 136:2261-2265.
82. Manners DJ, Meyer MT. The molecular structure of some glucans from the cell walls of *Schizosaccharomyces pombe*, Carbhydr Res 1977; 57:189-203.
83. Horisberger M, Rouvet-Vauthey M. Cell wall architecture of the fission yeast *Schizosaccharomyces pombe*. Experientia 1985; 41:748-750.
84. Kopecká M, Fleet GH, Phaff HJ. Ultrastructure of the cell wall of *Schizosaccharomyces pombe* following treatment with various glucanases. J Struc Biol 1995; 114:140-152.
85. Kornfeld R, Kornfeld S. Assembly of aspargine-linked oligosaccharides. Annu Rev Biochem 1985; 54:631-664.

86. Bush DA, Horisberger M, Horman I et al. The wall structure of *Schizosaccharomyces pombe*. J Gen Microbiol 1974; 81:199-206.
87. Moreno S, Ruíz T, Sánchez Y et al. Subcellular localization and glycoprotein nature of the invertase from the fission yeast *Schizosaccharomyces pombe*. Arch Microbiol 1985; 142:370-374.
88. Schweingruber AM, Schoenholzer F, Keller L et al. Glycosylation and secretion of acid phosphatase in *Schizosaccharomyces pombe*. Eur J Biochem 1986; 158:133-140.
89. Moreno S, Sánchez Y, Rodríguez L. Purification and characterization of the invertase from *Schizosaccharomyces pombe*. Biochem J 1990; 267:697-702.
90. Dibenedetto G, Cozzani I. Nonspecific acid phosphatase EC-3.1.3.2 from *Schizosaccharomyces pombe*. Purification and physical chemical properties. Biochemistry 1975; 14:2847-2852.
91. Ballou C, Ballou L, Ball G. *Schizosaccharomyces pombe* glycosylation mutant with altered cell surface properties. Proc Natl Acad Sci USA 1994; 91:9327-9331.
92. Chappell TG, Hajibagheri MAN, Asyscough K et al. Localization of an α 1,2 galactosyltransferase activity to the Golgi apparatus of *Schizosaccharomyces pombe*. Mol Biol Cell 1994; 5:519-528.
93. Smith DG, Svoboda A. Golgi apparatus in normal cells and protoplasts of *Schizosaccharomyces pombe*. Microbios 1972; 5:177-182.
94. Chappell TG, Warren G. A galactosyltransferase from the fission yeast *Schizosaccharomyces pombe*. J Cell Biol 1989; 109:2693-2702.
95. Kendall RL, Yamada R, Bradshaw RA. Cotranslational amino-terminal processing. Method Enzymol 1990; 185:398-407.
96. Huang S, Elliott RC, Liu PS et al. Specificity of cotranslational amino-terminal processing of proteins in yeast. Biochemistry 1987; 26:8242-8246.
97. Moerschell RP, Hosokawa Y, Tsunasawa S et al. The specificities of yeast methionine aminopeptidase and acetylation of amino-terminal methionine *in vivo*. J Biol Chem 1990; 265:19638-19643.
98. Schafer WR, Trueblood CE, Yang CC et al. Enzymatic coupling of cholesterol intermediates to a mating pheromone precursor and to the ras protein. Science 1990; 249:1133-1139.
99. Giannakouros T, Armstrong J, Magee AI. Protein prenylation in *Schizosaccharomyces pombe*. FEBS Lett 1992; 297:103-106.
100. James G, Olson EN. Fatty acylated proteins as components of intracellular signaling pathways Biochemistry 1990; 29:2623-2634.
101. Sambucetti LC, Schaber M, Kramer R et al. The *fos* gene product undergoes extensive post-translational modification in eukaryotic but not in prokaryotic cells. Gene 1986; 43:69-77.
102. Miyamoto C, Chizzomte R, Crowl R et al. Molecular cloning and regulated expression of the human *c-myc* gene in *E. coli* and *Saccharomyces cerevisiae*: comparison of the protein products. Proc Natl Acad Sci USA 1985; 82:7232-7236.
103. Jakubowicz T, Cytrynska M, Kowalczyk W et al. Phosphorylation of acidic ribosomal proteins by ribosome-associated protein kinases of

Saccharomyces cerevisiae and *Schizosaccharomyces pombe*. Acta Biochimica Polonica 1993; 40:497-505.

104. Koegl M, Courtneidge SA, Superti-Furga G. Structural requirements for the efficient regulation of the Src protein tyrosine kinase by Csk. Oncogene 1995; 11:2317-2329.
105. Romanos MA, Scorer CA, Clare JJ. Foreign gene expression in yeast: a Review. Yeast 1992; 423-488.
106. Suaréz-Renduele P, Villa L, Arbesú MJ et al. The proteolytic system of the fission yeast *Schizosaccharomyces pombe*. FEMS Microbiology Letters 1991; 81:215-220.
107. Tabuchi M, Iwaihara O, Ohtani Y et al. Vacuolar protein sorting in fission yeast: cloning. biosynthesis transport and processing of carboxypeptidase Y from *Schizosaccaromyces pombe*. J Bacteriol 1997; 179(13):in press.
108. Uritani M, Muramatsu T, Yoshino K et al. Alteration of cellular components induced by nutrition shift-down in *Schizosaccharomyces pombe*. Yeast 1990; 6(special issue):S136
109. Satoh S, Suzuki H, Widyastuti U. Identification and characterization of genes induced during sexual differentiation in *Schizosaccharomyces pombe*. Current Genetics 1994; 26:31-37.
110. Takegawa K, Dewald DB, Emr SC. *Schizosaccharomyces pombe* Vps-34p, a phosphatidylinositol-specific PI 3-kinase essential for normal cell growth and vacuole morphology. J Cell Sci 1995; 108:3745-3756.
111. Davey J, Davis K, Imai Y et al. Isolation and characterization of Krp, a dibasic endopeptidase required for cell viability in the fission yeast *Schizosaccharomyces pombe*. EMBO 1994; 13:5910-5921.
112. Imai Y, Yamamoto M. *Schizosaccharomyces pombe sxa* 1^+ and *sxa* 2^+ encode putative protease involved in the mating response. Mol Cell Biol 1992; 12:1827-1834.
113. Ladds G, Rasmussen M, Davey J. Characterization of Sxa 2, a protease involved in pheromone communication in fission yeast. Biochemical Society Transactions 1995; 23:565S.
114. Arbesu JM, Valle E, Saurez-Renduelles P. Purification and characterization of aminopeptidase YspI from *Schizosaccharomyces pombe*. Yeast 1993; 9:637-644.
115. Villa L, Suaretz-Rendueles P. Dipeptidyl aminopeptidase yspI mutants of *Schizosaccharomyces pombe*: Genetic mapping of *dpa1*$^+$ on chromosome III. FEMS Microbiology Lett 1994; 120:211-216.
116. Xu HP, White M, Marcus S et al. Concerted action of RAS and G proteins in the sexual response pathways of *Schizosaccharomyces pombe*. Mol Cell Biol 1994; 14:50-58.
117. Obara T, Nakafuku M, Kaziro Y. Isolation and characterization of a gene encoding a G protein α subunit from *Schizosaccaromyces pombe* involvemnt in mating and sporulation pathways. Proc Natl Acad Sci USA 1991; 88:5877-5881.
118. Gallo GJ, Scheutz TJ, Kingston RE. Regulation of heat shock factor in *Schizosaccharomyces pombe* more closely resembles regulation in mammals than in *Saccharomyces cerevisiae*. Mol Cell Biol 1991; 11:281-288.

119. Gallo GJ, Prentice H, Kingston RE. Heat shock factor is required for growth at normal temperature in the fission yeast *Schizosaccharomyces pombe.* Mol Cell Biol 1993; 13:749-761.
120. Shatzman AR. Expression systems. In: Davis J, Rosenberg M, eds. Current Opinion in Biotechnology. Current Biology Ltd UK, 1993:4:517-615.
121. Cirino G, Peers SH, Flower RJ et al. Human recombinant lipocortin I has acute local anti-inflammatory properties in the rat paw edema test. Proc Natl Acad Sci USA 1989; 86:3428-3432.
122. Cregg JM, Vedvick TS, Raschke WC. Recent advances in the expression of foreign genes in *Pichia pastoris.* Bio/Technology 1993; 11:905-910.
123. Janowicz ZA, Merckelbach A, Eckart M et al. Expression system based on the methylotrophic yeast *Hansenula polymoropha.* Yeast 1988; 4:S155.
124. Fleer R, Yeh P, Amellal N et al. Stable multicopy vectors for high-level secretion of recombinant human serum albumin by *Kluyveromyces* yeasts. Bio/Technology 1991; 9:968-975.
125. Heslot H, Nicaud J-M, Fabre E et al. Cloning of the alkaline extracellular protease gene of *Yarrowia lipolytica* and its use to express foreign genes. In: Nga BH, Lee YK eds. Microbiology Applications in Food Biotechnology. Amsterdam: Elsevier Science 1990:27-45.
126. Kondo K, Miura Y, Sone H et al. High-level expression of a sweet protein, monellin, in the food yeast *Candida utilis.* Nature Biotech 1997; 1:453-457.
127. Manna F, Del Giudice L, Schreil WH et al. Two extramitochondrial circular DNA species in the petite negative yeast *Schizosaccharomyces pombe*: relative abundance and size determination by electron microscopy. Curr Genet 1982; 5:187-189.
128. Broach JR. Construction of high copy number yeast vectors using 2 µm circle sequences. Meth Enzymol 1983; 101:307-325.
129. Yamamoto M. Fission yeast. In: Barr PJ, Brake AJ, Valenzuela P eds. Yeast Genetic Engineering. MA: Butterworth Publishers. 1989:53-64.
130. Losson R, Lacroute F. Plasmids carring the yeast OMP decarboxylase structural and regulatory genes:Transcription regulation in a foreign environment. Cell 1983; 32:371-377.
131. Leroy D, Baldin V, Ducommun, B. Characterization of an active GST-human cdc2 fusion protein kinase expressed in the fission yeast *Schizosaccharomyces pombe:* A new approach to the study of cell cycle control proteins. Yeast 1994; 10:1631-1638.
132. Bröker M, Bäuml O. New expression vectors for the fission yeast *Schizosaccharomyces pombe.* FEBS Lett 1989; 248:105-110.
133. Belsham GJ, Barker DG, Smith AE. Expression of polyoma virus middle-T antigen in *S. cerevisiae.* Eur J Biochem 1986; 156:413-421.
134. Pobjecky N, Rosenberg GH, Dinter-Gottlieb G et al. Expression of the β-glucuronidase gene under the control of the CaMV 35 S promoter in *Schizosaccharomyces pombe.* Mol Gen Genet 1990; 220: 314-316.

135. Atkins D, Izant JG. Expression and analysis of the green fluorescent gene in the fission yeast *Schizosaccharomyces pombe*. Curr Genet 1995; 28:585-588.
136. Zhao Y, Cao J, O'Gorman MRG et al. Effect of human immunodeficiency virus type 1 protein R (vpr) gene expression of basic cellular function of fission yeast *Schizosaccharomyces pombe*. J Virology 1996; 70:5821-5826.
137. Ficca AG, Testa L, Tocchini-Valentini GP. The human β2-adrenergic receptor expressed in *Schizosaccharomyces pombe* retains its pharmacological properties. FEBS Lett 1995; 377:140-144.
138. Sander P, Grünewald S, Reiländer H et al. Expression of the human D2s dopamine receptor in the yeasts *Saccharomyces cerevisiae* and *Schizosaccharomyces pombe*:A comparative study. FEBS Lett 1994; 344:41-46.
139. Arkinstall S, Edgerton M, Payton M et al. Co-expression of the neurokinin NK2 receptor and G protein components in the fission yeast *Schizosaccharomyces pombe*. FEBS Lett 1995; 375:183-187.
140. Ueda K, Shimabuku AM, Konishi H et al. Functional expression of human P-glycoprotein in *Schizosaccharomyces pombe*. FEBS Lett 1993; 330:279-282.
141. Jackson MR, Burchell B. Expression of human liver epoxide hydrolase in *Schizosaccharomyces pombe*. Biochem J 1988; 251:931-933.
142. Loddenkötter B, Kammerer B, Fischer K et al. Expression of the functional mature chroloplast triose phosphate translocator in yeast internal membranes and purification of the histidine-tagged protein by a single metal-affinity chromatography step. Proc Natl Acad Sci USA 1993; 90:2155-2159.
143. Sauer N, Caspari T, Klebl F et al. Functional expression of the *Chlorella* hexose transporter in *Schizosaccharomyces pombe*. Proc Natl Acad Sci USA 1990; 87:7949-7952.
144. Yamazaki S, Sato K, Suhara K et al. Importance of the Proline-rich region following signal-anchor sequence in the formation of correct conformation of microsomal cytochrome P450. J Biochem 1993; 114:652-657.
145. Smerdon GR, Aves SJ, Walton EF. Production of human gastric lipase in the fission yeast *Schizosaccharomyces pombe*. Gene 1995; 165:313-318.
146. Tokunaga M, Kawamura A, Yonekyu S et al. Secretion of mouse α-amylase on fission yeast *Schizosaccharomyces pombe*. Yeast 1993; 9:379-387.
147. Zárate V, Belda F. Characterization of the heterologous invertase produced by *Schizosaccharomyces pombe* from the *SUC2* gene of *Saccharomyces cerevisiae*. J Appl Bacteriol 1996; 80:45-52.
148. Sánchez Y, Moreno S, Rodríguez L. Synthesis of *Saccharomyces cerevisiae* invertase by *Schizosaccharomyces pombe*. FEBS Lett 1988; 234:95-99.
149. Sande S, Privalsky ML. Reconstitution of thyroid hormone receptor and retinoic acid receptor function in the fission yeast *Schizosaccharomyces pombe*. Molecular Endocrinology 1994; 8:1455-1464.

Chapter 2

Constitution of Expression Vectors

Hiromichi Kumagai

Introduction

Yeast can be grown rapidly and to high cell density on simple media compared with the growth of mammalian cells, and it can be easily manipulated, as is *E. coli*. Yeast-based gene expression systems have other benefits shown by the *E. coli* system, combined with the advantages of being a eukaryotic expression system, including post-translational modification and the high-level production of secreted proteins as well as soluble cytosolic proteins.[1]

The yeast *Schizosaccharomyces pombe* (fission yeast) is one of the most intensely studied and well characterized of the yeast species.[2,3] *S. pombe* may be more suitable than *Saccharomyces cerevisiae (S. cerevisiae)* for isolating human cDNA expression libraries because of its closer relationship with mammalian cells.[4,5] Previous analyses have suggested that the *S. pombe* expression system may be useful for the production of various mammalian proteins. Fission yeast is nonpathogenic, as is *S. cerevisiae*, and has attracted attention as a means of deacidification of grape musts in the alcohol industry, because it can metabolize malic acid. Since *S. pombe* provides an appropriate host for heterologous gene expression, it is very important to develop a new type of *S. pombe* expression vector. The use of *S. pombe* for protein production, however, has been limited because the development of a host-vector system for this yeast has lagged behind that of *S. cerevisiae* for many years.

The vector in *S. pombe* consists of a bacterial origin of replication, selectable marker in *E. coli*, yeast selection marker and an autonomous replication sequence (ARS; *S. pombe arsı*[6,7] or *S. cerevisiae*

Foreign Gene Expression in Fission Yeast Schizosaccharomyces pombe, edited by Yuko Giga-Hama and Hiromichi Kumagai.

2 μm *ori*[8,9] are available). Most *S. pombe* expression vectors have been based on *S. pombe ars1*-containing plasmids and a promoter and terminator for the efficient transcription of the foreign gene. These components should be selected for the construction of efficient expression vectors for each protein production. The recent *S. pombe* molecular genetics studies have led to a greater understanding of the components of each expression vector.

In this chapter, we review the expression vectors now available to the subsequent chapters. The method of obtaining a high level of gene expression using *S. pombe* as a host are described here.

Vector

The strategy for the construction of efficient expression vectors in *S. pombe* is to combine the transcriptional initiation and termination signals active in *S. pombe* (promoter and terminator) with selection markers for *E. coli* and *S. pombe*, in addition to DNA elements, which function as the replication origin. In *S. pombe*, as an autonomous replication sequence (ARS), *S. pombe ars1*[6,7] and the *S. cerevisiae* 2 μm *ori*[8,9] are available. Most *S. pombe* expression vectors contain ARS derived from the *S. cerevisiae* 2 μm *ori* or from the *S. pombe ars1*.[10-16] The recent findings regarding the molecular genetics of *S. pombe* have clarified the components needed for replication.

Chromosomal integration offers a more stable alternative for extrachromosmal maintenance of foreign DNA. Many integrating vectors currently available for use in *S. pombe*, which do not contain ARS, are based on the complementation of *S. pombe* mutations with *S. cerevisiae* genes, including *leu1* and *ura4* genes.[17] Genomic DNA blot analysis showed that the transformants have one or more copies of the plasmid vector integrated via homologous recombination. In *S. pombe*, integration by homologous recombination is usually more frequent than by nonhomologous recombination.[18] For certain loci, homologous recombination may represent only about 5-10% of the integration events.

Table 2.1 is a summary of the fission yeast expression vectors.

Selection Marker and Transformation

The efficiency of transformation is generally high regardless of the *S. pombe* strain used. Protoplast,[19] electroporation[20] and lithium acetate[21,22] procedures were developed for the transformation of *S. pombe*. Transformation efficiencies of between 10^3 to 10^5 are possible using each procedure. The protoplast method for the transformation of *S. pombe* involves enzymatic removal of the cell wall to produce spheroplasts which could take up DNA on treatment with calcium and polyethylene glycol. Transformants were then plated in a selective, isotonic agar for regeneration of the cell wall. The transformation frequency is about 10^4/μg DNA. This efficiency can be increased by using lipofectin. The lithium acetate method gives a higher frequency of transformation compared with the protoplast procedure: the transformation frequency is more than 10^5 transformants/μg of DNA. The third approach for transformation, the electroporation procedure, has been used to obtain high efficiency (more than 10^5/μg DNA).

The most commonly used markers for the selection of transformants after transformation[23-25] have been the *S. cerevisiae* genes *LEU2* and *URA3*, which complement the *S. pombe leu1* and *ura4*, respectively. The corresponding genes from *S. pombe* are also used as selective markers. The commonly used cloning markers are summarized in Table 2.2.[23-29] Dominant selection is useful because the selection markers increase the range of host strains that can be tested, especially in rich medium. The aminoglycoside antibiotic G418,[30] hygromycin,[31] chloramphenicol, phleomycin and bleomycin[32] resistances have been demonstrated. The G418-resistance marker (encoded by the *E. coli Tn903* transposon) can be used for the selection because most strains of *S. pombe* are sensitive to G418, although

Table 2.1. Selectable markers commonly used for* S. pombe *transformation

Marker	Origin	Comments
CAN1	*S. cerevisiae*	complements *S. pombe can1* (arginine permease gene)
LEU2	*S. cerevisiae*	complements *S. pombe leu2*
URA3	*S. cerevisiae*	complements *S. pombe ura4*
ade1	*S. pombe*	complements *ade1*
ade6	*S. pombe*	complements *ade6*
arg3	*S. pombe*	complements *arg3*
his3	*S. pombe*	complements *his3*
his7	*S. pombe*	complements *his7*
leu1	*S. pombe*	complements *leu1*
ura4	*S. pombe*	complements *ura4*
Tn903 kan^r	*E. coli*	selection using G418
Hyg^r	*Streptomyces hygroscopicus*	selection using hygromycin

the sensitivity to the G418 is dependent on the culture medium used (unpublished data).

Multiple copies of the G418 resistance marker are needed to confer resistance in *S. pombe*. Tohda et al[33] demonstrated an efficient expression of heterologous genes using the pTL2M1 vector, which contains G418 resistance marker combined with the *LEU2* (chapter 4). The copy number of pTL2M1 vectors in a cell increases dramatically in the presence of G418, because only the recombinant yeasts containing a high copy number of the vectors are viable in a culture medium containing a high concentration of G418.

Replication Origin

Most extrachromosomal replicons are based on plasmids containing the *S. pombe ars1* or the *S. cerevisiae* 2 µm *ori*.[6-9] Although even a bacterial plasmid such as pBR322 was reported to replicate in *S. pombe*, the transformation frequency using this plasmid was very low. The addition of the *S. pombe ars1* sequence or the *S. cerevisiae* 2 µm *ori* leads to a high frequency of transformation and a reduction in the copy number of the vector in *S. pombe*. Plasmid vectors containing *ars1* or 2 µm *ori*, however, are unstable mitotically, resulting in the accumulation of plasmid-free cells without selection. Plasmid vectors based on 2 µm are mitotically unstable (more than 30% loss per generation without selection), and their copy number is low (5-10 copies). In contrast, plasmids based on *S. pombe ars1* are present in multiple copies per cell (15 to 80), but are mitotically unstable. The *S. pombe*-derived sequence *stb* yields stable transfomants both mitotically and meiotically. These vectors[6] containing *ars1* and *stb* have a copy number of about 80 and are lost at a rate of 13% per generation without selection.[21]

An ultra-high copy number vector[33,34] has been developed by a combination of the G418 selection and *LEU* selection using *ars1* and *stb* sequences, as described in chapter 4. The use of the selectable marker G418 resistance gene resulted in a very high copy number (more than 200 copies per cell in some cases) and the vector was very stable.

The ARS elements in *S. pombe* have been identified and were shown to colocalize with origins on chromosomes.[35-37] Various fragments that mediate autonomous replication, as judged by high transformation frequency and plasmid instability, have been reported. The identification of ARS binding proteins[38] should define the ARS functions. Among them, an effective replication origin should be selected to ensure that the vector is present and stable in *S. pombe*.

Table 2.2. Schizosaccharomyces pombe *expression vector systems*

Vector	Replication Origin	Promoter	Yeast markers	Comments
pART	*arsl*	*adh* promoter	*LEU2*	
pEVP11	2 μm *ori*	*adh* promoter	*LEU2*	
pCHY21	*arsl*	*fbp1* promoter	*URA3*	regulated by glucose
pREP	*arsl*	*nmtl* promoter	*LEU2*	regulated by thiamine
pTL2M1/pAL7	*arsl*	hCMV promoter	*neo*^{r}, *LEU2/URA3*	high copy number(>200)
pSL2M1/pAL7	*arsl*	hCMV promoter	*neo*^{r}, *LEU2/URA3*	high copy number(>200), P-factor signal sequence
pSLF	*arsl*	CaMV promoter	*LEU2*	
pSLF101	*arsl*	CaMV-tet	*LEU2*	regulated by tetracyclin
pSM1/2	2 μm *ori*	SV40 promoter	*LEU2*	
pART1/N795	*arsl*	*adh* promoter	*LEU2*	regulated by glucocorticoid hormone (the hormone receptor is co-expressed)

Promoter

Many different promoters have been used to successfully direct the expression of foreign genes in *S. pombe*. Yeast promoters consist of three elements which regulate the initiation of transcription (including upstream activation sequences), TATA elements and initiator elements. The optimization of these three elements are necessary for the construction of an efficient expression vector.

The promoters available for *S. pombe*[39-52] are described below. The well-studied promoter is derived from *S. pombe* genes encoding abundant glycolytic enzymes, alcohol dehydrogenase(ADH). pEVP11 including the *adh* promoter were available to express heterologous genes efficiently (chapters 5 and 6). The *S. cerevisiae* promoters generally function poorly, although the *S. cerevisiae ADH1* promoter was used to produce antithrombin III and the respective promoters produce glycosidases of *S. cerevisiae*.

In contrast, there are many mammalian derived promoters active in *S. pombe*, including Simian virus 40 (SV40) early, human cytomegalovirus (hCMV), human chorionic gonadotropin α-subunit, adenovirus region 3, HIV-1 LTR promoters and HSP TATA elements. These promoters suggest that *S. pombe* and mammals may be similar in certain aspects of the regulatory mechanism of various cellular functions. The SV40 early promoter has also been commonly used to produce heterologous proteins in *S. pombe*. The hCMV promoter has been reported to show the strongest promoter activity, which is 10 times higher than that of the SV40 early promoter.[43] The other constitutive promoters are Cauliflower mosaic virus (CaMV) 35S promoter[48] and the tomato nitrate reductase *nia* gene promoter.[49]

Some of the promoters are regulated to some extent, but the promoter activities are poorly regulated. This makes them undesirable for use in a large-scale culture. Powerful tightly regulated promoters are sometimes needed for the production of toxic proteins, for example.

Maundrell[50] recently described the *nmt1* promoter, which is repressed by thiamine, and developed a regulated expression vector in *S. pombe* based on the thiamine-repressible *nmt1* gene. Basi et al[12] modified the TATA box in the *nmt1* promoter to attenuate the activity. As described in detail in chapter 3, the *nmt1* promoter-based expression vectors called pREP1 and pREP3 and their derivatives have been constructed for the insertion of cDNA in a multicloning site. These vectors have been successfully used to express various heterologous genes, as described in chapter 3. The regulatory element thiamine is a growth regulator which affects the transcription of a variety of genes in *S. pombe*. It may not be possible for the *nmt1* promoter-based expression vector to produce a protein in large quantities, because the depletion of the thiamine in the medium is needed.

Other examples of inducible promoters are the *fbp1* promoter[51] and *S. pombe* invertase promoter (Kaoru Takegawa, Kagawa University: personal communication), which are regulated by glucose concentration. The promoter activity of the *fbp1* gene is repressed over a range of greater than 100-fold by glucose, which also affects a variety of metabolic processes in the cell. Moreover, this promoter is derepressed in the stationary phase. Some examples of protein production using a vector containing the *fbp1* promoter are summarized in chapter 10.

The other regulatory system is the use of the *Tn10*-encoded Tet repressor-operator system to control transcription in *S. pombe*. Faryar and Gatz[52] reported the expression vector containing the plant caulioflower mosaic virus 35S promoter combined with the Tet regulatory elements, which respond to the inducer tetracycline.

The mammalian glucocorticoid response elements are used to drive expression in *S. pombe*.[10] This glucocorticoid regulated system is induced up to 70-fold by glucocorticoid hormones when the hormone receptor is co-expressed.

There are few comparative data on the promoter activities in fission yeast, because the expression experiments were carried out using different expression vectors. Forsburg[53] recently reported comparative data on promoter activities which should allow a direct comparison of activity when a new type of promoter is characterized in the future. However, it is difficult to evaluate the efficiency of each expression vector based only on their promoter activities. The length between TATA elements and the initiation site providing the best expression level have been reported to be 40 to 120 bp in *S. cerevisiae* and around 50 bp in *S. pombe.*[44]

Yeast transcriptional terminators are usually present in expression vectors for the mRNA 3' end formation. To obtain the maximal expression level, a termination sequence should be inserted. The transcriptional termination of yeast mRNA is less well understood than that in bacteria. The yeast mRNAs follow a pattern of termination, processing and polyadenylation of pre-mRNA similar to that of higher eukaryotes. The SV40 terminator or a terminator from the human lipocortin I gene, for example, is used in chapter 4.

Secretion

In the past decade, heterologous gene expression has been treated extensively investigated and many expression systems have been explored. The secretion of foreign proteins is often necessary for their correct folding and biological activity. Moreover, when the secretion system is used to produce a protein, it is highly advantageous for purification from protein-free culture medium. However, there are areas where improvement is needed, especially the signal sequences. In *S. pombe*, the analysis of the secretion signal has lagged behind that of other yeasts, including *S. cerevisiae.*

Many pharmaceutical-related proteins including cytokines are naturally secreted and can adopt their correct conformation by folding within the secretory pathway. Secretory proteins must enter into the secretory pathway (which is mainly composed of endoplasmic reticulum (ER) and the Golgi apparatus) and the removal of the signal sequence of each protein by signal peptidase in host cells must occur for the correct folding, the formation of disulfide bonds, glycosylation and other aspects of post-translational modification during the secretion pathway.

The secretion of heterologous proteins which are important commercially from *S. cerevisiae* and *Pichia pastoris* has been successful.[1] In *S. pombe*, the secretion of foreign proteins is also an important area, because like higher eukaryotes, *S. pombe* is known to contain galactosyltransferase.[54,55] Other yeasts such as *Pichia pastoris* and *S. cerevisiae* do not contain the transferase. Further, secretion systems are useful for the production of correctly-folded and naturally-secreted proteins. In addition, when the protein of interest is unstable or toxic, a secretion system may be used. It is easier to purify the proteins produced using a secretion system.

Protein secretion in yeasts is directed by an amino-terminal signal sequence which mediates cotranslational translocation into the ER, as in mammalian cells. During the secretion pathway, the signal sequence is cleaved by a signal peptidase, and glycosyl structures may be added to the lumen of the ER. Glycosylated proteins are then transported into the Golgi apparatus, in which the modification of the glycosylated moieties occurs. The development of convenient secretion vectors is important for the production of various proteins with natural structures. A few mammalian proteins including mouse α-amylase,[56] human placental alkaline phosphatase (chapter 10) and anti-thrombin III[39] have been reported to be produced from *S. pombe* in a secreted form.

Heterologous proteins may be secreted from yeast using either a foreign or a yeast signal. The signal sequence of human placental alkaline phosphatase works well in *S. pombe*. Although the foreign signals, including that of the alkaline phosphatase, could work as efficiently as those from yeast, this is often not the case. The secretion of foreign proteins from *S. pombe* is a relatively unexplored area of potential interest. Recent reports[57] concluded that the essential core for the signal sequence is 6-15 hydrophobic amino acids, which may be interrupted by non-hydrophobic residues.

Using a signal sequence derived from the *S. pombe* mating pheromone, P-factor,[58] heterologous proteins may be secreted efficiently from *S. pombe*, as described in chapter 11.

Future Perspectives

The fission yeast *Schizosaccharomyces pombe* is the most intensely studied and well characterized of the yeast species, aside from *S. cerevisiae*. *S. pombe* may be more suitable than *S. cerevisiae* for isolating human cDNA expression libraries because of its closer relationship with mammalian cells, and *S. pombe* is proving increasingly attractive as host for producing various kinds of industrial proteins.

There have recently been several successes in the production of therapeutic proteins from yeast, including the recombinant subunit vaccine against hepatitis B virus, human proinsulin, granulocyte-macrophage colony stimulating factor (GM-CSF), human serum albumin (HSA) and so on. In the food industry, yeast is used for the production of chymosin and glucoamylase. The development of expression systems is becoming increasingly important with the increasing industrial use of target proteins. Although a foreign gene may be successfully inserted into the vector, a high level expression of the foreign gene may not be obtained in some cases. Gene expression is a complex process, comprised of the replication of the vector, transcription, translation, mRNA stability, protein stability and so on. Further analysis of the multi-step process of gene expression is needed to obtain an optimum level of foreign gene expression.

Finally, the safety aspect should be considered when an expression system in *S. pombe* is used. Since vectors free of bacterial DNA may be advantageous in foreign gene expression for the food and drug industries, the bacterial-derived genes in an expression vector must sometimes be removed.

Schizosaccharomyces pombe is an attractive host organism for the expression of heterologous genes. Recent advances in molecular biology will enable the development of much more efficient expression vectors, including a secretion signal and an inducible promoter.

References

1. Romanos MA, Scorer SA, Clare JJ. Foreign gene expression in yeast: a review. Yeast 1992; 8:423-488.
2. Moreno S, Klar A, Nurse P. Molecular genetic analysis of fission yeast *Schizosaccharomyces pombe*. In: Guthrie C, Fink GR eds. Methods in Enzymology. Volume 194: Guide to yeast genetics and molecular biology. San Diego: Academic Press Inc. 1991:795-823.
3. Hayles J, Nurse P. Genetics of the fission yeast *Schizosaccharomyces pombe*. Ann Rev Genetics 1992; 26:373-402.
4. Fukui Y, Kaziro Y. Molecular cloning and sequence analysis of a *ras* gene from *Schizosaccharomyces pombe*. EMBO J 1985; 4:687-691.
5. Lee MG, Nurse P. Complementation used to clone a human homologue of the fission yeast cell cycle control gene *cdc2*. Nature 1987; 327:31-35.
6. Losson R, Lacroute F. Plasmids carrying the yeast OMP decarboxylase structural and regulatory genes: Transcriptional regulation in a foreign environment. Cell 1983; 32:371-377.
7. Maundrell K, Hutchison A, Shall S. Sequence analysis of ARS elements in fission yeast. EMBO J 1988; 7:2203-2209.
8. Gaillardin C, Fournier P, Budar F et al. Replication and recombination of 2μ DNA in *Schizosaccharomyces pombe*. Curr Genet 1983; 7:245-253.
9. Wright A, Maundrell K, Heyer W-D et al. Vectors for the construction of gene banks and the integration of cloned genes in *Schizosaccharomyces pombe* and *Saccharomyces cerevisiae*. Plasmid 1986; 15:156-158.

10. Picard D, Schena M, Yamamoto KR. An inducible expression vector for both fission and budding yeast. Gene 1990; 86:257-261.
11. Russel P. Gene cloning and expression in fission yeast. In: Nasim A, Young P, Johnson BF, eds. Molecular Biology of the Fission Yeast. San Diego:Academic Press, Inc, 1989; 243-271.
12. Basi G, Schmid E, Maundrell K. TATA box mutations in the *Schizosaccharomyces pombe nmt1* promoter affect transcription efficiency but not the transcription start point or thiamine repressibility. Gene 1993; 123:131-136.
13. Faryar K, Gatz C. Construction of a tetracycline-inducible promoter in *Schizosaccharomyces pombe.* Curr Genet 1992; 21:345-349.
14. Hoffman CS, Winston F. Glucose repression of transcription of the *Schizosaccharomyces pombe fbp1* gene occurs by a cAMP signaling pathway. Genes Dev 1991; 5:561-571.
15. Maundrell K. *nmt1* of fission yeast: A highly transcribed gene completely repressed by thiamine. J Biol Chem 1990; 265:10857-10864.
16. McLeod M, Stein M, Beach D. The product of the *mei3+* gene expressed under control of the mating type locus, induces meiosis and sporulation in fission yeast. EMBO J 1987; 6:729-736.
17. Grimm C, Kohli J, Murray J et al. Genetic engineering of *Schizosaccharomyces pombe*: A system for gene disruption and replacement using the *ura4* gene as a selectable marker. Mol Gen Genet 1988; 215:81-86.
18. Grallert B, Nurse P, Patterson TE. A study of integrative transformation in *Schizosaccharomyces pombe.* Mol Gen Genet 1993; 238:26-32.
19. Beach D, Nurse P. High frequency transformation of the fission yeast *Schizosaccharomyces pombe.* Nature 1981; 290:140-142.
20. Hood MT, Stachow C. Transformation of *Schizosaccharomyces pombe* by electroporation. Nucleic Acid Res 1990; 18:688.
21. Heyer WD, Sipiczki M, Kholi. Replication plasmids in *Schizosaccharomyces pombe*: Improvement of symmetric segregation by a new genetic element. Mol Cell Biol 1986; 6:80-89.
22. Okazaki K, Okazaki N, Kume K et al. High-frequency transformation method and library transducing vectors for cloning mammalian cDNAs by trans-complementation of *Schizosaccharomyces pombe.* Nucleic Acid Res 1990; 18:6485-6489.
23. Yamamoto M. Fission yeast. In: Barr PJ, Brake AJ, Valenzuela P, eds. Yeast Genetic Engineering. Boston:Butterworth. 1989; 53-64.
24. Kikuchi Y, Kitazawa Y, Shimatake H et al. The primary structure of the *leu1* gene of *Schizosaccharomyces pombe.* Curr Genet 1988; 14:375-379.
25. Bach M-L. Cloning and expression of the OMP decarboxylase gene *ura4* from the yeast *Schizosaccharomyces pombe.* Curr Genet 1987; 12:527-534.
26. Burke JD, Gould KL. Molecular cloning and characterization of the *Schizosaccharomyces pombe his3* gene for use as a selectable marker. Mol Gen Genet 1994; 242:169-176.

27. Waddell S, Jenkins JR. *arg3+*, a new selection marker system for *Schizosaccharomyces pombe*: Application of *ura4+* as a removable integration marker. Nucleic Acids Res 1995; 23:1836-1837.
28. Cottarel G. The *S. cerevisiae HIS3* and *LYS2* genes that complement *S. pombe his5-303* and *lys1-131* mutations, respectively: New selectable markers and new multi-purpose multicopy shuttle vectors, pSP3 and pSP4. Curr Genet 1995; 28:380-383.
29. Apolinario E, Nocero M, Jin M et al. Cloning and manipulation of the *Schizosaccharomyces pombe his7+* gene as a new selectable marker for molecular genetic studies. Curr Genet 1993; 24:491-495.
30. Kudla B, Persuy M-A, Gaillardin C et al. Construction of an expression vector for the fission yeast *Schizosaccharomyces pombe*. Nucleic Acid Res 1988; 16:8603-8617.
31. Burland TG, Pallotta D,Tardif MC et al. Fission yeast promoter-probe based on hygromycin resistance. Gene 1991; 100:241-245.
32. Sakai K, Sakaguchi J, Yamamoto M. High frequency cotransformation by copolymerisation of plasmids in the fission yeast *Schizosaccharomyces pombe*. Mol Cell Biol 1984; 4:651-656.
33. Tohda H, Okada H, Giga-Hama Y et al. A copy number controlled expression vector for the fission yeast *Schizosaccharomyces pombe* Gene 1994; 150:275-280.
34. Giga-Hama Y, Tohda H, Okada H et al. High-level expression of human lipocortin I in the fission yeast *Schizosaccharomyces pombe* using a novel expression vector. Bio/Technology 1994; 1:400-404.
35. Caddle MS, Calos MP. Specific initiation at an origin of replication from *Schizosaccharomyces pombe*. Mol Cell Biol 1994; 14:1796-1805.
36. Dubey DD, Zhu J, Carlson DL et al. Three ARS elements contribute to the *ura4* replication origin region in the fission yeast, *Schizosaccharomyces pombe*. EMBO J 1994; 13:3638-3647.
37. Wohlgemuth JG, Bulboaca GH, Moghadam M et al. Physical mapping of origins of replication in *Schizosaccharomyces pombe*. Mol Biol Cell 1994; 5:839-849.
38. Murakami Y, Huberman JA, Hurwitz J. Identification, purification and molecular cloning of autonomously replicating sequence-binding protein 1 from fission yeast, *Schizosaccharomyces pombe*. Proc Natl Acad Sci USA 1996; 93:502-507.
39. Bröker M, Ragg H, Karges HE. Expression of human antithrombin III in *Saccharomyces cerevisiae* and *Schizosaccharomyces pombe*. Biochim Biophys Acta 1987; 908:203-213.
40. Russell PR. Transcription of the triose-phosphate isomerase gene of *Schizosaccharomyces pombe* initiates from a start point different from that in *Saccharomyces cerevisiae*. Nature 1983; 301:167-169.
41. Kuranda MJ, Robbins PW. Cloning and heterologous expression of glycosidase genes from *Saccharomyces cerevisiae*. Proc Natl Acad Sci USA 1987; 84:2585-2589.
42. Jones R, Moreno S, Nurse P et al. Expression of the SV40 promoter in fission yeast: identification and characterization of an AP1-like factor. Cell 1988; 53:659-667.

43. Toyama R, Okayama H. Human chorionic gonadotropin α and human cytomegalovirus promoters are extremely active in the fission yeast *Schizosaccharomyces pombe*. FEBS Letts 1990; 268:217-221.
44. Swaminathan S, Malhotra P, Manohar CF et al. Activation of a dual adenovirus promoter containing nonconsensus TATA motifs in *Schizosaccharomyces pombe*: Role of TATA sequences in the efficiency of transcription. Nucleic Acid Res 1993; 21:2737-2746.
45. Toyama R, Bende SM, Dhar R. Transcriptional activity of the human immunodeficiency virus-1 LTR promoter in fission yeast *Schizosaccharomyces pombe*. Nucleic Acid Res 1992; 20:2591-2596.
46. Prentice HL, Kingston RE. Mammalian promoter element function in the fission yeast *Schizosaccharomyces pombe*. Nucleic Acid Res 1992; 20:3383-3390.
47. Russell PR, Hall BD. The primary structure of the alcohol dehydrogenase gene from the fission yeast *Schizosaccharomyces pombe*. J Biol Chem 1983; 258:143-149.
48. Pobjecky N, Rosenberg GH, Dinter-Gottlieb G et al. Expression of the β-glucuronidase gene under the control of the CaMV 35S promoter in *Schizosaccharomyces pombe*. Mol Gen Genet 1990; 220:314-316.
49. Truong H-N, Caboche M, Daniel-Vedele F. The tomato *nia* gene promoter functions in fission yeast but not in budding yeast Plant. Mol Biol 1992; 20:361-364.
50. Maundrell K. *nmt1* of fission yeast. J Biol Chem 1990; 265: 10857-10864.
51. Hoffman CS, Winston FA. Transcriptionally regulated expression vector for the fission yeast *Schizosaccharomyces pombe*. Gene 1989; 84:473-479.
52. Faryar K, Gatz C. Construction of a tetracycline-inducible promoter in *Schizosaccahromyces pombe*. Curr Genet 1992; 21:345-349.
53. Forsburg SL. Comparison of *Schizosaccharomyces pombe* expression systems. Nucleic Acid Res 1993; 21:2955-2956
54. Moreno S, Ruiz T, Sanchez Y et al. Subcellular localization and glycoprotein nature of the invertase from the fission yeast *Schizosaccharomyces pombe*. Arch Microbiol 1985; 142:370-374.
55. Dibenedetto G, Cozzani I. Non-specific acid phosphatase from *Schizosaccharomyces pombe*. Purification and physical chemical properties. Biochemistry 1975; 14:2847-2852.
56. Tokunaga M, Kawamura A, Yoneyasu S et al. Secretion of mouse α-amylase from fission yeast *Schizosaccharomyces pombe*: The presence of chymostatin-sensitive protease activity in the culture medium. Yeast 1993; 9:379-387.
57. Murakami Y, Huberman JA, Hurwitz J. Identification, purification, and molecular cloning of autonomously replicating sequence-binding protein 1 from fission yeast, *Schizosaccharomyces pombe*. Proc Natl Acad Sci USA 1996; 4:502-507.
58. Imai Y, Yamamoto Y.The fission yeast mating pheromone P-factor: Its molecular structure, gene structure and ability to induce gene expression and G1 arrest in the mating partner. Genes and Dev 1994; 8:328-338.

Part II

Development of Expression Vectors

CHAPTER 3

Exploiting the Thiamine Biosynthetic Gene *nmt1* for Regulated Expression of Heterologous DNAs in Fission Yeast

Kinsey Maundrell

Introduction

Ectopic expression of cloned DNA sequences provides a general and widely used approach for investigating gene function. When the host is a genetically tractable microorganism such as yeast, where the genetic background can be modified virtually at will, and the full power of yeast molecular genetics can be brought to bear, ectopic expression offers immense scope for analysis of complex biological problems.

The obvious prerequisite for the success of this approach is the availability of suitable vectors for heterologous expression, and given that many expressed sequences have biological activity which can interfere with host cell function, the ideal expression system is one which can be downregulated effectively. In the budding yeast *Saccharomyces cerevisiae*, several inducible genes have been well characterized over the years and the promoters of these genes have provided the basis for constructing a range of inducible expression vectors. Among the most widely used are *GAL1/GAL10* and *CUP1* promoters which are activated by growth in galactose and copper

Foreign Gene Expression in Fission Yeast Schizosaccharomyces pombe, edited by Yuko Giga-Hama and Hiromichi Kumagai.

respectively, the *PHO5* promoter regulated by inorganic phosphate levels and the *MET3* promoter which is activated by growth in medium lacking methionine (reviewed in ref. 1). Each of these promoters has its particular attributes and this has provided a wide choice of vectors for heterologous expression studies.

The development of comparable vectors for the fission yeast *Schizosaccharomyces pombe*, lagged behind the budding yeast for many years. Basic research on the organism had failed to identify suitable transcriptionally regulated genes and none of the inducible promoters described for budding yeast functioned satisfactorily. Expression studies in fission yeast were thus restricted to using strong constitutively active promoters such as the alcohol dehydrogenase (*adh*)[2] or certain viral promoters[3,4] and this imposed serious limitations to working with this organism. In recent years therefore a number of laboratories have addressed the need for inducible heterologous expression and four different systems have now been reported. Two of these achieve regulated expression by exploiting promoters of the recently described endogenous genes *fbp1*[5] and *nmt1*[6] which are repressed by glucose and thiamine respectively; and two exploit mechanisms imported from other organisms in which transcription is controlled by exogenously added ligand, either glucocorticoid hormone[7] or tetracycline.[8]

Of the four systems described to date, expression based on *fbp1* is probably the least convenient in practice because the promoter becomes derepressed in low glucose as cells enter stationary phase. The two systems activated by addition of small molecules are highly regulatable and have the advantage that in both cases the ligands are unrecognized by the host organism and consequently are without pleiotropic effects on yeast physiology. Furthermore, both can be adapted for use in fission and budding yeasts, which is frequently an advantage in practice. The only minor inconvenience with these systems is that both require co-expression of an appropriate 'effector' molecule, (nuclear receptor or tet repressor) in order to couple the presence of ligand to transcriptional activation, and probably for this reason neither system has yet been widely adopted.

To date, by far the most frequently used vectors for regulated expression in fission yeast have been those based on the thiamine-repressible *nmt1* gene and this will form the subject of the current review. The *nmt1* promoter is highly active under inducing conditions and furthermore, because cells are able to grow equally well either with or without thiamine, the induction conditions per se do not significantly disturb the cellular metabolism. In this chapter I will first describe what is known about the control of thiamine biosynthesis in fission yeast, since this provides a background for the underlying physiology of this system. I will then describe the expanding family of *nmt1*-based expression vectors, and finally, I will discuss some of the practical issues of concern when using these vectors for heterologous expression studies.

Thiamine Biosynthetic Pathway in *S. pombe*

Thiamine pyrophosphate is a universal and essential co-factor for the activity of several key enzymes involved in carbohydrate metabolism. In yeast cells, optimal growth requires elevated levels of the vitamin. In the wild, this is achieved mostly through an avid transport machinery which concentrates the vitamin from the environment[9,10] while in cells growing under thiamine-free conditions, key enzymes of the thiamine biosynthetic pathway are induced and the high intracellular concentration of thiamine necessary for rapid growth is maintained by de novo synthesis. Even so, in cells growing in defined minimal medium without thiamine, the intracellular thiamine concentration achievable by the endogenous pathway is apparently the rate limiting factor for growth since addition of exogenous thiamine to the culture medium causes an increase in culture doubling time of 5-10% (unpublished observations). Thus we might anticipate that key enzymes which catalyze the rate-limiting steps in thiamine biosynthesis would be highly expressed and indeed this turns out to be the case.

Thiamine consists of two constituent groups, namely a pyrimidine, 4-amino-5-hydroxy-methyl-2-methylpyrimidine and a thiazole, 5-(2-hydroxyethyl)-4-methylthiazole. In the budding yeast where the pathway has been studied most extensively these two elements are synthesized in separate pathways as phosphorylated intermediates which are subsequently combined to yield thiamine monophosphate. A phosphatase completes the synthesis of thiamine. Thiamine pyrophosphate, the active form as a co-factor in carbohydrate metabolism, is synthesized from thiamine by direct transfer of pyrophosphate from ATP.[11]

The available evidence suggests that the biosynthetic pathway in fission yeast is similar to that described in budding yeast.[12] Genetic analysis to date has identified mutants at four loci, *thi1* to *thi4*, which result in thiamine auxotrophy.[10] More detailed analysis of the mutant phenotypes indicates that *thi2* encodes an enzyme involved in synthesis of the thiazole component, *thi3* is required for synthesis of the pyrimidine, and *thi4* plays a role in thiamine biosynthesis following completion of the pyrimidine and thiazole groups.[10] *thi1*, also known as *ntf1*[13] (see below) is a positively acting transcription factor essential for the expression of all three other components.[14] In an independent series of experiments, *thi3* and *thi2* were molecularly cloned and sequenced and were referred to as *nmt1*[6] and *nmt2*[15] respectively (see below).

Transcriptional Control of Thiamine Biosynthetic Genes

The first gene shown to be transcriptionally regulated by thiamine in fission yeast was *pho4*, a membrane-bound acid phosphatase which is present in the periplasmic space and which is most probably required to dephosphorylate thiamine mono- and di-phosphates prior to uptake of free thiamine into the cell.[16] Expression of *pho4* was shown to be downregulated by thiamine; however, as a candidate gene for constructing regulated expression vectors, *pho4* was not ideal because it was found to be only partially repressed by exogenous thiamine. A search specifically to identify genes whose expression is more tightly regulated by thiamine was therefore undertaken using a differential hybridization approach in which a *S. pombe* genomic library was screened sequentially with radioactive cDNA probes synthesized using template RNA isolated from cultures grown either in the presence or absence of thiamine. This led to the identification of two genes which were shown by subsequent Northern blotting and nuclear 'run on' analysis to be highly expressed in cells grown in the absence of thiamine and totally repressed when thiamine was added.[6,15] These two genes were named *nmt1* and *nmt2* to denote the fact that in both cases there was *no message* in *thiamine*. A study of the kinetics of induction and repression indicated that these two genes were coordinately regulated in response to thiamine, and an analysis of the promoters of the two genes was undertaken in order to search for common motifs which might underlie the thiamine-mediated control of transcription. Resection experiments of the *nmt1* promoter generated a series of deletions towards the ATG initiator codon and analysis of these mutants showed that all of the information necessary for transcriptional repression by thiamine was present within the 243 bp upstream of the ATG.[17] The 230 bp fragment upstream of the *nmt2* coding sequence which was present in the original genomic fragment was also able to confer thiamine repressibility[15] and a homology search between these two promoter sequences revealed the perfectly conserved TATA box sequence, 5' TATATAAA, and a second sequence, 5' CATCAATTGAA, which occurred the same distance downstream of the TATA box in the two promoters. Transcript mapping showed that the mRNA of both genes initiated within this motif. The sequence was not found in the promoters of any other genes of the fission yeast, suggesting that it might play a role in thiamine regulation. Somewhat surprisingly however, substitution of this sequence in the *nmt1* promoter with the sequence 5' ACGAAAGGTCC, in which only the central A residues at the site of transcription initiation is retained, did not affect either the level of transcription nor thiamine repressibility,[17] so the role of this element and its potential as a recognition site for transcription factors remains to be elucidated.

A genetic analysis of the factors which regulate transcription of the thiamine repressible genes has been initiated.[14] This has identified three negatively acting *tnr* genes—so-called because mutations in these genes lead to a *t*hiamine *n*on-*r*epressible phenotype in which all the known genes normally repressed by thiamine (*nmt1/thi2*, *nmt2/thi3*, *pho4* and *thi4*) are constitutively expressed; and one positively acting transcription factor encoded by the *thi1* gene. The *thi1* gene was also identified serendipitously as a high copy number suppressor of a *nmt1:wee1* induced phenotype and was referred to as *ntf1*.[13] *thi1/ntf1* is a Cys6 zinc-finger type transcription factor required for full expression of the thiamine sensitive genes,[13,18] although it is of interest that in a *thi1* deleted strain transcription of these genes is reduced but not abolished[13] ("Practical considerations", below). None of the promoter binding sites for any of these activities has been determined nor is it known how thiamine acts to control transcription. In view of the similarities in transcriptional control mechanisms which have been noted between fission yeast and higher eukaryotes,[4,19,20] more detailed investigation of these questions could provide a fruitful direction for future research.

Construction of pREP Vectors

Construction of the REP vectors was based on the *nmt1* regulatory sequences (Fig. 3.1A) and has been described in detail elsewhere.[21] Briefly the *nmt1* coding sequence was replaced by a multiple cloning site (MCS) and the expression cassette consisting of a 1.2 kb region upstream of the *nmt1* gene together with a 1 kb fragment downstream containing the polyadenylation site was inserted into the bacterial plasmid pUC119.[22] Early versions of the REP plasmids (pREP1, pREP2) carried the *Nde*I restriction site (CA'TA_TG) at the promoter ATG, which was fashionable for a time since it allows direct fusion of the promoter to the coding sequence of interest. However, in practice this is seldom useful for heterologous expression since it usually requires mutagenesis of the incoming cDNA to create an appropriate *Nde*I site and even when this is done, the desirability of this site is questionable since the sequence 5' CAT upstream of the initiator codon is not optimal for efficient translation.[23] Alternative versions of the MCS were also constructed which carried the blunt end *Msc*I site (TGG'CCA) immediately upstream of the ATG initiator codon (REP3, REP4) and this has proved more generally useful. In the meantime, however, several laboratories including ours (see Fig. 3.1B and footnote to Table 3.1) have produced more versatile cloning polylinkers.

Table 3.1. Characteristics of the pREP family of vectors

vector	size (kb)	MCS[1]	promoter[2]	selection[3]
pREP1	8.8	N	+++	*LEU2*
pREP2	8.3	N	+++	*ura4*
pREP3	8.8	M	+++	*LEU2*
pREP3B	8.6	S	+++	*leu1*
pREP4	8.3	M	+++	*ura4*
pREP4B	8.4	S	+++	*ura4*
pREP5	9.2	M	+++	*ade6*
pREP7	8.1	M	+++	*his5*
pREP43	8.8	M	++	*LEU2*
pREP44	8.3	M	++	*ura4*
pREP83	8.8	M	+	*LEU2*
pREP84	8.3	M	+	*ura4*

1 Nucleotide sequences of the alternative multiple cloning sites are as follows:
N gttaaat CATA**TG**TCGACTCTAGAGGATCCCCGGG **taa**aagg
(*Nde*I/*Sal*I/(*Xba*I)/*BamHI*/*Sma*I)
M gttaaat TGGCC**ATG**TCGACTCTAGAGGATCCCCGGG **taa**aagg
(*Msc*I/*Sal*I/(*Xba*I)/*BamHI*/*Sma*I)
S gttaaat TGGCCACTAGTAACGGCCGCCAGTGTGCTGGAATTC
TGCAGATATCCATCACACTGGCGGCCGCTCGACTCTAGAG
GATCCCCGGG **taa**aagg
(*Msc*I/*Spe*I/*Eag*I/(*BstX*I)/(*EcoR*I)/(*Pst*I)/*EcoR*V/(*BstX*I)/*Eag*I/*Not*I/(*Xba*I)/*BamH*I/*Sma*I)
nmt1 encoded sequences are shown in lower case. The initiator and stop codons are shown in bold. Restriction sites in brackets are not unique.
2 +++ indicates the full strength wild type *nmt1* promoter;
++ indicates the weaker promoter in which the TATA box is mutated to ATAAA;
+ indicates the weakest promoter in which the TATA box is deleted.
3 *LEU2* is from *S. cerevisiae*, all other genes are from *S. pombe*.

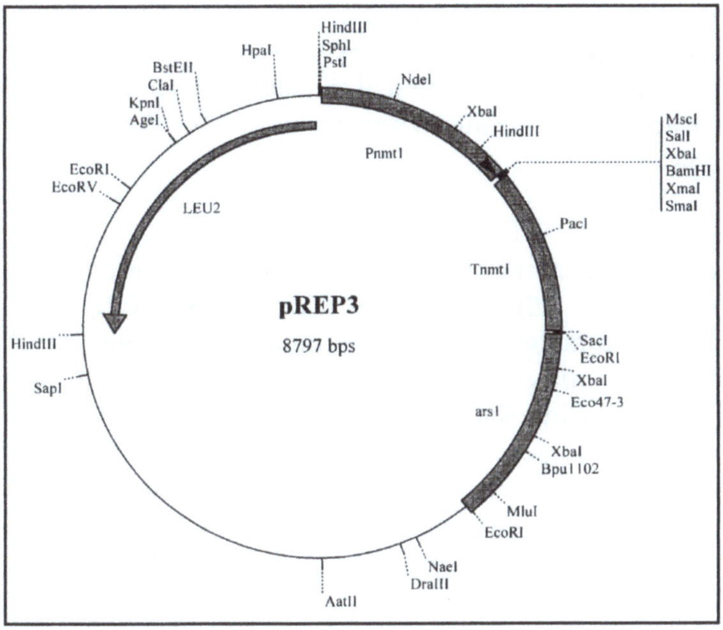

Fig. 3.1. General structure of the pREP vectors.
(A) The structure of pREP3 which has the *LEU2* selectable marker and the '*Msc*I' polylinker is shown. The locations of key restriction sites are indicated. pREP4, pREP5 and pREP7, which carry the *ura4*, *ade6* and *his5* selectable markers respectively, have the same basic structure.
(B) The structure of pREP3B which has the *S. pombe leu1* selectable marker and an extended polylinker is shown. For integration by homologous recombination at the marker locus, pREP3B can be linearized at the sites indicated. For pREP4B, which has the *ura4* marker, the corresponding sites are *Avr*II, *Stu*I, *Bbs*I and *Bsu*36.

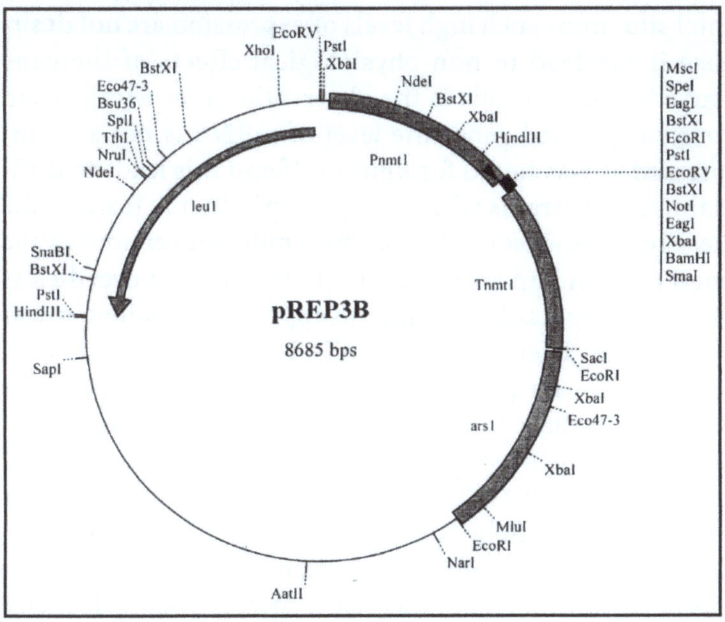

This basic *nmt1* expression cassette has been used as a starting point for constructing a family of related vectors differing either in the genetic marker used for selection of transformants or in the level of promoter activity. These alternative forms of the vector are described below.

Selectable Markers

Three selectable markers were employed initially: a 2.2 kb sequence encoding the *LEU2* gene from *S. cerevisiae* (REP1, REP3) which is known to complement the *leu1*-32 mutation in fission yeast;[24] the 1.8 kb *S. pombe ura4* gene (REP2, REP4) and the 2.6 kb *Sca*I-*Spe*I fragment encompassing the *S. pombe ade6* gene (REP5). Subsequently the cloning of *S. pombe his5* (ref. 25; Maundrell K and Leberrer E, unpublished results) provided a fourth selectable marker which was linked to and expressed using the *adh* promoter (pREP7). All selectable markers were cloned into the *Hin*dIII site of pUC119 such that transcription is away from the MCS. More recently the *S. pombe leu1* gene has also been used as a selectable marker since the homology with the chromosomal gene is useful for targetted plasmid integration[26] ("pRIP integrating vectors", below).

Promoter Strength

The fully induced *nmt1* promoter is probably one of the most active promoters in the cell[6] and has provided the strongest expression system currently available for fission yeast.[27] Clearly, in many experimental situations such high levels of expression are not desirable because it can lead to non-physiological effects of the gene being expressed. Added to this is the observation that with certain heterologously expressed genes the level of mRNA is not as completely repressed as was found for *nmt1* itself and this leads to difficulty in maintaining strains where the gene product is toxic to the cell. For these reasons we sought to produce mutated versions of the *nmt1* promoter which had reduced activity but which nevertheless retained thiamine repressibility, and the approach we adopted was to make progressive deletions of the TATA box with the aim of reducing the overall transcriptional efficiency.[28] Quantitative estimates of promoter activity based on expression of the chloramphenicol acetyl transferase (CAT) reporter gene indicated that mutation of the original sequence 5' TATATAAA to 5' ATAAA or simply to 5' AT reduced promoter strength by approximately one and two orders of magnitude, respectively.[28] Moreover, in addition to reducing the induced levels of expression, there was also a corresponding decrease in the residual activity detectable under repressing conditions. These two promoter variants have now formed the basis of two further series of vectors referred to as the REP40 series (ATAAA) and REP80 series (AT). Experiments using β-galactosidase as the reporter gene have largely confirmed the original estimates of promoter activity, although some differences were noted, most likely due to differences in the stability of the reporter gene product.[27] The reduced level of expression and the reduction in residual repressed transcription characteristic of these promoter mutants has also been confirmed experimentally.[29]

pRIP Integrating Vectors

pREP vectors contain the *S. pombe ars1* fragment cloned into the *Eco*RI site of the pUC119 polylinker; this favors extrachromosomal replication of multiple copies of the plasmid and reduces the frequency sequence rearrangement. We have now subcloned many genes into these vectors and our accumulated experience has been that, particularly in the case of genes encoding a toxic gene product, expression from extrachromosomally replicating plasmids frequently results in wide phenotypic variability. This is due in part to cell-to-cell variation in plasmid copy number, but also to an increased frequency of mutations in the sequence present in high copy number, and thus to an increased likelihood of producing a cell with an inactive version of the toxic gene product. Increasingly therefore where possible we have tended to work with single copy chromosomal integrants of the construct under investigation, and this has consistently produced a more stable and homogeneous phenotype and thus a more reliable strain for subsequent analysis.

For the purpose of chromosomal integration the pRIP series of plasmids were made by simply deleting the *ars1* element. Non-homologous integration can occur in fission yeast; however, the frequency is low compared to homologous integration.[26] Additionally, we felt the need for more systematic integration of plasmids; therefore several changes were introduced into the pREP plasmids to facilitate targeted integration. Based on the studies of Keeney and Boeke[26] the principal changes in this regard were to make use of the homologous *leu1* gene instead of *S. cerevisiae LEU2*, and to exploit more fully the unique restriction sites within each of the selectable markers for linearizing plasmids before transformation (Fig. 3.1.B). The characteristics of the vectors currrently in use are described in Table 3.1.

Practical Considerations[31]

Criteria for an ideal regulatable expression vector are rapid induction, high stimulation above background, low residual expression when repressed, and non-stressful conditions for derepression. The strengths and weaknesses of the pREP vectors with regard to these criteria are discussed below.

The activity of the thiamine repressible *nmt1* promoter is controlled by the concentration of thiamine *inside* the cell. In cells growing in defined medium without thiamine, the biosynthetic machinery maintains a level of around 10 pmoles/10^7 cells.[31] This is well below the threshold for transcriptional repression, which we have estimated to be around 50 pmoles/10^7 cells.[31] Upon addition of thiamine to a log phase culture growing in minimal medium, an efficient transport system actively concentrates the vitamin from the medium, resulting in a rapid increase in the internal thiamine concentration. If the thiamine in the medium is present in excess (> 4 μm; see Fig. 1 of ref. 31), the intracellular thiamine concentration can increase by up to 1000-fold the basal level, to a maximum of around 10,000 pmoles/10^7 cells before the transport system shuts down. If on the other hand cells are transferred to medium in which the thiamine concentration is limiting, the intracellular thiamine concentration rises until the medium is fully depleted and this results in an internal thiamine concentration which is therefore less than the maximum possible. In either case once cells are washed and allowed to resume normal growth in thiamine free medium the internal thiamine concentration decreases by approximately 50% every 3 h, a rate of decrease which suggests that the vitamin is not itself metabolized as the cell grows but is simply diluted out as the cell mass increases. During this process, transcription of the thiamine biosynthetic genes is repressed almost immediately once cells are exposed to thiamine, and is not reactivated until the concentration inside the cell falls to below a threshold of about 50 pmoles/10^7cells.[31]

The physiology of thiamine response as outlined above has a number of implications for using the *nmt1* promoter to express heterologous genes. Firstly, repression by thiamine is for all practical purposes all-or-nothing since the internal thiamine concentration rapidly rises beyond the threshold required for transcriptional repression.[31] Thus it is not possible to regulate the level of expression by varying the amount of thiamine added to the medium. The effect of varying the external thiamine concentration is instead to alter the lag time prior to induction, since as described above this will affect the initial concentration of thiamine attained within the cell and hence the length of time taken before the internal concentration falls to the critical point where the *nmt1* promoter is derepressed. The other parameter which it is important to consider in terms of reproducibility of the induction response is cell density at the time of thiamine addition. Clearly, since thiamine is scavenged so efficiently from the external medium, if the supply of extracellular thiamine becomes limiting, the maximal level attained inside each cell will decrease as cell number increases and this again will influence the lag period prior to induction. The innoculation density of the initial overnight culture preceeding an experiment is therefore an important parameter.

A standard protocol for induction of *nmt1*-mediated expression of heterologous genes is as follows:

day 1 Innoculate a 10 ml overnight culture at about 10^6cells/ml in minimal medium (plus appropriate supplements) containing 4 mM thiamine. Grow at 30°C without shaking for 24 hr. By this time cells should have entered early stationary phase.

day 2 Wash cells and resuspend at a concentration of 2-5 x 10^5 cells/ml in minimal medium without thiamine. Grow the culture at 30°C in a shake flask for 18-20 hr. The final density of the culture should be about 5 x 10^6-10^7 cells/ml. If for some reason growth has been inhibited and the final density is lower, it is unlikely that induction will be optimal for the reasons given above.

day 3 Harvest the cells.

Typically, following this standard protocol, mRNA expression begins at about 12 hr post thiamine withdrawal and is maximal by 16 hr. Curiously mRNA levels usually decline to a somewhat lower (probably steady state) level by around 25-30 hr. Recombinant protein levels normally peak at about 18-20 hr. Thus despite the lengthy lag period which precedes induction, during which the internal thiamine concentration falls to below the critical threshold, the induction itself is reasonably rapid considering the high levels of mRNA which are ultimately achieved. Furthermore, provided the factors above are kept constant, the induction response is highly reproducible.

As mentioned earlier, it is not possible to achieve variation in expression levels by adjusting the external vitamin levels, but two possibilities currently exist for manipulating promoter activity. One is to exploit the differences in promoter strength of the TATA box mutants described above (see subsection "Promoter strength"). The other is to make use of the strain deleted for the positively acting transcription factor *nft1*. As reported recently, expression levels of *nmt1* in such a strain are considerably attenuated yet the strain remains viable in the absence of thiamine and furthermore, repression by thiamine is unaffected.[13] This strain therefore offers a convenient way of investigating the effects of reduced expression of a given protein without the need for subcloning into a different vector.

Finally, since the full strength *nmt1* promoter is particularly powerful it is possible to drive expression of several cDNAs from different pREP vectors in the same cell. By using all of the available selectable markers we have routinely co-transformed cells with up to four different cDNAs in a single transformation.[32-34] In our experience not all transformants express high levels of each of the recombinant proteins; nevertheless, where antibodies against each of the proteins have been available it has been relatively straightforward to select transformants which express appropriate levels of each protein. Moreover, once isolated, a given strain seems to maintain the same pattern of expression of each of the recombinant proteins during subsequent storage and subculturing. Thus transformation with multiple cDNA constructs provides a valuable means of studying the interaction in vivo between recombinant proteins of interest.

In conclusion, the fission yeast is proving to be an attractive and amenable host organism in which to study problems of eukaryotic cell and molecular biology. Many of the powerful molecular genetic procedures developed for *S. cerevisiae* can be applied directly to *S. pombe*, and this, coupled with the development of several regulatable expression systems such as the one described above, has now made fission yeast an attractive option for both academic and industrial research.

References

1. Romanos MA, Scorer CA, Clare JJ. Foreign gene expression in yeast: a review. Yeast 1992; 8:423-88.
2. Russell PR, Nurse P. cdc25$^+$ functions as an inducer in the mitotic control of fission yeast. Cell 1986; 45:145-53.
3. Gmunder H, Kohli J. Cauliflower mosaic virus promoters direct the efficient expression of a bacterial G418 resistance gene in *Schizosaccharomyces pombe*. Molec Gen Genet 1989; 220:95-101.
4. Jones R, Moreno S, Nurse P et al. Expression of the SV40 promoter in fission yeast: identification and characterisation of an AP1-like factor. Cell 1988; 53:659-67.
5. Hoffman CS, Winston F. A transcriptionally regulated expression vector for the fission yeast *Schizosaccharomyces pombe*. Gene 1989; 84:473-479.

6. Maundrell K. *nmt1* of fission yeast. A highly transcribed gene completely repressed by thiamine. J Biol Chem 1990; 265:10857-64.
7. Picard D, Schena M, Yamamoto KR. An inducible expression vector for both fission yeast and budding yeast. Gene 1990; 86:257-61.
8. Faryar K, Gatz C. Construction of a tetracycline-inducible promoter in *Schizosaccharomyces pombe*. Curr Genet 1992; 21:345-9.
9. Iwashima A, Nishino H, Nose Y. Carrier mediated transport of thiamine in baker's yeast. Biochim Biophys Acta 1973; 330:222-34.
10. Schweingruber AM, Dlugonski J, Edenharter E, et al. Thiamine in *Schizosaccharomyces pombe:* dephosphorylation, intracellular pool biosynthesis and transport. Curr Genet 1991; 19:249-54.
11. Leder IG. Thiamine, biosynthesis and function. In: Greenberg DM, ed. Metabolic Pathways. Metabolism of Sulphur Compounds. New York: Academic Press, 1975; 3(VII):57-85.
12. Praekelt UM, Byrne KL, Meacock PA. Regulation of THI4(MOL1), a thiamine biosynthetic gene of *Saccharomyces cerevisiae*. Yeast 1994; 10:481-90.
13. Tang CSL, Bueno A, Russell P. *ntf1*$^{+}$ encodes a 6-cysteine zinc finger-containing transcription factor that regulates the *nmt1* promoter in fission yeast. J Biol Chem 1994; 269:11921-6.
14. Schweingruber AM, Fankhauser H, Dlugonski J et al. Isolation and characterization of regulatory mutants from *Schizosaccharomyces pombe* invloved in thiamine regulated gene expression. Genetics 1992; 130:445-9.
15. Manetti AGO, Rosetto M, Maundrell KG. *nmt2* of fission yeast: a second thiamine repressible gene co-ordinately regulated with *nmt1*. Yeast 1994; 10:1075-82.
16. Schweingruber ME, Fluri R, Maundrell K, et al. Identification and characterization of thiamin repressible acid phosphatase in yeast. J Biol Chem 1986; 261:15877-82.
17. Rosetto M. Tesi di Laurea. Universita Degli Studi di Siena. Siena, Italy 1991.
18. Fankhauser H, Schweingruber ME. Thiamine repressible genes in *Schizosaccharomyces pombe* are regulated by a Cys_6 zinc-finger motif-containing protein. Gene 1994; 147:141-4.
19. Russell P, Nurse P. *Schizosaccharomyces pombe* and *Saccharomyces cerevisiae*: a look at yeasts divided. Cell 1986; 45:781-2.
20. Prabhala G, Rosenberg GH, Kaufer NF. Architectural features of premRNA introns in the fission yeast *Schizosaccharomyces pombe*. Yeast 1992; 8:171-82.
21. Maundrell K. Construction of the pREP and pRIP thiamine repressible expression vectors for fission yeast. Gene 1993:127-30.
22. Vieira J, Messing J. Production of single-stranded plasmid DNA. Meth Enzymol 1987; 153:3-11
23. Kozak M. The scanning model for translation: an update. J Cell Biol 1989; 108:229-241.
24. Moreno S, Klar A, Nurse P. Molecular genetic analysis of fission yeast *Schizosaccharomyces pombe*. Methods Enzymol 1991; 194: 795-823.

25. Erickson FL, Hannig EM. Characterization of *Schizosaccharomyces pombe his1* and *his5* cDNAs. Yeast 1995; 11:157-67.
26. Keeney JB, Boeke JD. Efficient targetted integration at *leu1*-32 and *ura4*-294 in *Schizosaccharomyces pombe*. Genetics 1994; 136:849-56.
27. Forsburg SL. Comparison of *Schizosaccharomyces pombe* expression systems. Nucl Acids Res 1993; 21:2955-56.
28. Basi G, Schmid E, Maundrell K. TATA box mutations in the *nmt1* promoter of *S. pombe* affect transcriptional efficiency but not the site of transcription initiation nor the repressibility by thiamine. Gene 1993; 123:131-6.
29. Labib K, Craven RA, Crawford K et al. Dominant mutants identify ew roles for $p34^{cdc2}$ in mitosis. EMBO J 1995; 14:2155-65.
30. Maundrell K, Huchinson A, Shall S. Sequence analysis of *ars* elements in fission yeast. EMBO J 1988; 7:2203-9.
31. Tommasino M, Maundrell K. Uptake of thiamine by *Schizosaccharomyces pombe* and its effect as a transcriptional regulator of thiamine sensitive genes. Curr Genet 1991; 20:63-66.
32. Arkinstall S, Payton M, Maundrell K. Activation of phospholipase Cg in *Schizosaccharomyces pombe* by co-expression of receptor and non-receptor tyrosine kinases. Mol Cell Biol 1995; 15:1431-8.
33. Arkinstall S, Edgerton M, Payton M et al. Co-expression of the neurokinin NK2 receptor and G protein components in the fission yeast *Schizosaccharomyces pombe*. FEBS Lett 1995; 375:183-7.
34. Arkinstall S, Maundrell K. A negative regulatory function for the protein tyrosine phosphatase PTP2C revealed by reconstruction of platelet-derived growth factor receptor signalling in *Schizosaccharomyces pombe*. Submitted.

CHAPTER 4

Development of Constitutive Expression Vector for the Highest Production Level of Heterologous Proteins in Fission Yeast *Schizosaccharomyces Pombe*

Hideki Tohda

Introduction

Heterologous protein production in the fission yeast *Schizosaccharomyces pombe* has been limited compared with the budding yeast *Saccharomyces cerevisiae*, although *S. pombe* has been intensively investigated in terms of molecular and cellular biology. The development of a host-vector system in *S. pombe* for the high-level production of heterologous proteins has lagged far behind that of such a system in *S. cerevisiae*.

To obtain the highest production level of heterologous protein using a plasmid vector, its stability during mitosis is most important. The *ars1* element[1] is commonly used as the source of the autonomous replication sequence in *S. pombe*.

Several promoters cloned from *S. pombe* or *S. cerevisiae* have been used for foreign gene expression in *S. pombe*. The alcohol dehydrogenase (*adh*) promoter[2] from *S. pombe* shows high transcriptional activity and has been utilized for parts of expression vectors such as pEVP11 and pART1.[3] Unlike *S. cerevisiae*, the mammalian promoter and its viral promoter elements such as the SV40 early

Foreign Gene Expression in Fission Yeast Schizosaccharomyces pombe, edited by Yuko Giga-Hama and Hiromichi Kumagai.

promoter[4] are functional in *S. pombe*. Toyama and Okayama reported[5] that human promoters of cytomegalovirus (hCMV) and chorionic gonadotropin α showed high transcriptional activity in *S. pombe* and initiated their transcriptions at the same sites as in mammalian cells. Both promoters were about 10-fold as active as the SV40 early promoter in a chloramphenicol acetyltransferase (CAT) assay. These activities were also higher than that of the *S. pombe adh* promoter.

To improve the mitotic stability of the expression vector, we constructed a "copy number-controlled" expression vector[6,7] in which an antibiotic resistance gene is regulated by a weak promoter, SV40. Using this expression cassette, the growth of the cells with a low copy number of the plasmid vector was prevented. The copy number was controlled by adding G418 to the culture medium, since the neomycin resistance-encoding gene (*neo*) was selected as the antibiotic-resistance gene. The transcription level of a foreign gene was maintained using a strong promoter, hCMV; thus, this vector carries two mammalian promoters, SV40 early promoter as the weak promoter and hCMV promoter as the strong promoter.

To improve the translation efficiency, we optimized the length of the 5'-untranslated region (*5'-UTR*) of the hCMV promoter, and the expression levels of the foreign genes were markedly improved.[7]

In this chapter, these two improvements in the expression vector (the copy number-controlled vector system and the optimization of the *5'-UTR*) are described. Human lipocortin I cDNA[8] and green fluorescent protein (GFP) cDNA from the bioluminescent jellyfish *Aequorea victoria*[9] were employed as foreign genes, and the CAT-encoding gene (*cat*) was used as the reporter to estimate the level of expression.

Copy Number-Controlled Expression by *neo*/G418 System

To obtain a high production level of the target protein, we constructed the copy number-controlled expression system using the *neo* transcriptional unit and the presence of G418 in the culture medium, with the *cat* gene as a reporter.

Structure of the cat *Expression Vector, pcD4-CAT*

Plasmid pcD4-CAT (Fig. 4.1a) is based on the Okayama-Berg pcD vector.[10] This plasmid contains the *neo* gene (indicated by Nm^R in the figure), regulated by the SV40 early promoter region, which is the same as the "*neo* transcriptional unit" described by Chen and

Okayama.[11] Plasmid pcD4-CAT also contains the *cat* gene driven by the hCMV promoter region, which is a part of the major IE gene of hCMV.[12]

Transformation of S. pombe *with pcD4-CAT*

S. pombe (*leu1-32 h⁻* strain) was cotransformed with pcD4 (no insert control) or pcD4-CAT by the lithium acetate method,[13] together with *Pst*I-linearized transducing vector pAL7,[13] which contains *LEU2* of *S. cerevisiae*. The transducing vector also contains the *ars1* and the *stb*[14] elements for mitotic stability. After the transformation procedure, cells were spread on an MMA[13] minimum plate and incubated at 32°C for 3 days. The colonies that appeared were transferred

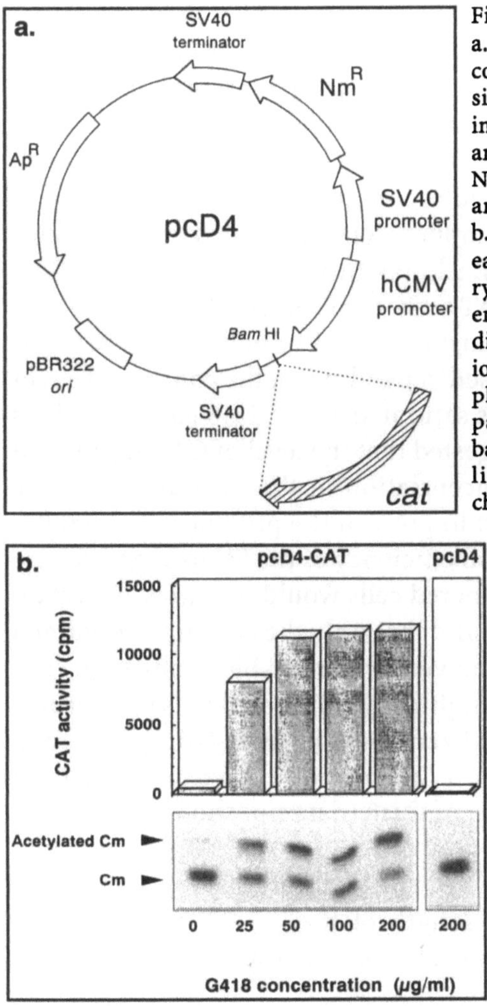

Fig. 4.1.
a. Structure of plasmid pcD4-CAT, a recombinant vector for *cat* gene expression. The *cat* gene (hatched arrow) was inserted into the *Bam*HI site of pcD4 [7] and was driven by the hCMV promoter. Nm^R: neomycin resistance gene, Ap^R: ampicillin resistance gene.
b. Comparison of the CAT activity of each cell-free lysate of pcD4-CAT carrying recombinants cultivated in different concentrations of G418 in the medium. Lower panel shows the autoradiogram of the thin layer chromatography of the reaction mixture. Upper panel shows the radioactivity of the bands of acetylated Cm measured by a liquid scintillation counter. Cm: chloramphenicol.

to a YE plate (0.5% yeast extract (Difco, Detroit, MI), 3% dextrose, 2% agar) containing 25 µg/ml of G418 (sulfate, Lifetechnologies Inc., Gaithersburg, MD) and cultivated at the same temperature for 3 to 5 days. With this screening method, recombinant yeast carrying *LEU2* and *neo* was cloned. The plasmids pcD4-CAT and pAL7 were thought to be rearranged to one molecule in yeast cells by homologous recombination.

The CAT Activity Depended on the G418 Concentration in the Medium

The *S. pombe* transformed with pcD4-CAT was grown in YPD medium (1% yeast extract (Difco), 2% peptone (Difco) 2% dextrose) in the presence of 0, 25, 50, 100 or 200 µg/ml of G418 at 32°C until the early stationary phase. After cell harvesting and breakage,[15] CAT assays were performed using 100 ng of extracted protein in a 50 µl reaction mixture containing 18.5 Bq of [^{14}C]chloramphenicol (2.11 GBq/mmol, Amersham International, Buckinghamshire, UK). Thin-layer chromatography was performed on Kieselgel 60F plates (Merck, Darmstadt, Germany). After autoradiography, the radioactivity of the bands of acetylated chloramphenicol was counted by liquid scintillation counter (LS5000TD, Beckman Instruments, Fullerton, CA).

Figure 4.1 b shows the CAT activity of cell-free lysates prepared from the transformants carrying pcD4-CAT or pcD4 (no insert control). The CAT activity increased depending on the concentration of G418 in YPD medium and was saturated at the G418 concentrations 50-200 µg/ml. This result suggested that the level of CAT expression is controlled by the G418 concentration in the medium. *S. pombe* requires the *neo* gene product to grow in the presence of G418. Because of the low transcriptional efficiency of the SV40 promoter, we suspected that low-copy-numbered cells would not grow in certain concentrations of G418. We found that only the cells that contained sufficient copies of recombinant vector were viable. Using high concentrations of G418, relatively high-copy-numbered cells were grown and the copy number was stabilized at a high-level. Using the strong hCMV promoter instead of the SV40 promoter, the expression level was saturated, even at 25 µg/ml of G418 (data not shown).

Optimization of the Length of the 5′-*UTR*

Next, we optimized the length of the 5′-*UTR* using human lipocortin I cDNA as a target gene. To construct six vectors that carry different lengths of the 5′-*UTR*, their expression levels were compared by the production of recombinant protein.

Construction of the Lipocortin I cDNA Expression Vector

The six plasmids pRL0L, pRL1L, pRL2L, pRL3L, pRL4L and pRL5L were constructed to determine the optimum length of the 5′-*UTR* located between the TATA box of the hCMV promoter and the start codon of human fibroblast lipocortin I cDNA. Lipocortin I cDNA was cloned from a human fibroblast cDNA library described elsewhere.[6] In all six plasmids, the 5′-*UTR* was based mainly on that of lipocortin cDNA. The length of the 5′-*UTR* was 360 nucleotides in pRL0L, 206 nucleotides in pRL1L, 56 nucleotides each in pRL2L and pRL3L, 42 nucleotides in pRL4L, and 28 nucleotides in pRL5L, as shown in Fig. 4.2a. The pRL2L and pRL3L plasmids have the same length of 5′-*UTR* except for the insertion of a *Sac*I site by mutation in pRL3L.

Transformation of* S. pombe *with Plasmids pRL0L-pRL5L

The *S. pombe leu1-32* h^- strain was cotransformed with plasmids pRL0L-pRL5L by the lithium acetate method, together with the *Pst*I-linearized transducing vector pAL7. Transformed cells were screened and cultivated by the same procedure described in section 4.2.2.

***Comparison of the Production Levels in the Various Lengths of 5′-*UTR**

Transformants carrying the plasmids pRL0L-pRL5L, respectively, were cultivated in YPD medium containing 200 µg/ml of G418 at 32°C. Cells were harvested and broken using glass beads in the denaturation buffer. The total soluble protein extracts (20 µg) were analyzed by 0.1% sodium dodecyl sulfate (SDS) - 7.5-15% polyacrylamide gel electrophoresis (PAGE) and stained with Coomassie Brilliant Blue (CBB).

Figure 4.2b shows the expression levels of lipocortin I in *S. pombe* carrying each plasmid. The highest levels were present in the transformants carrying pRL2L, pRL3L or pRL4L, with a 5′-*UTR* length of around 50 nucleotides. The lipocortin I level in strains with pRL5L was slightly lower than the highest level, which may be due to the translation initiating from an internal methionine (the third residue from initial methionine residue).

Comparison of the Amount of mRNA in the Transformants

To verify the increase of the translational level due to the difference of the length of the *5'-UTR*, Northern blots of two transformants (those carrying pRL1L and pRL2L) were done. Total RNA (20 µg) was prepared as follows. Yeast cells (the same cells as those described above) were treated with zymolyase-20T (Seikagaku-Kogyo, Tokyo, Japan) to get spheroplasts. After disruption in 4M guanidine thiocyanate, the RNA was extracted with phenol, electrophoresed on 2.2 M formaldehyde/1% agarose gels and transferred to

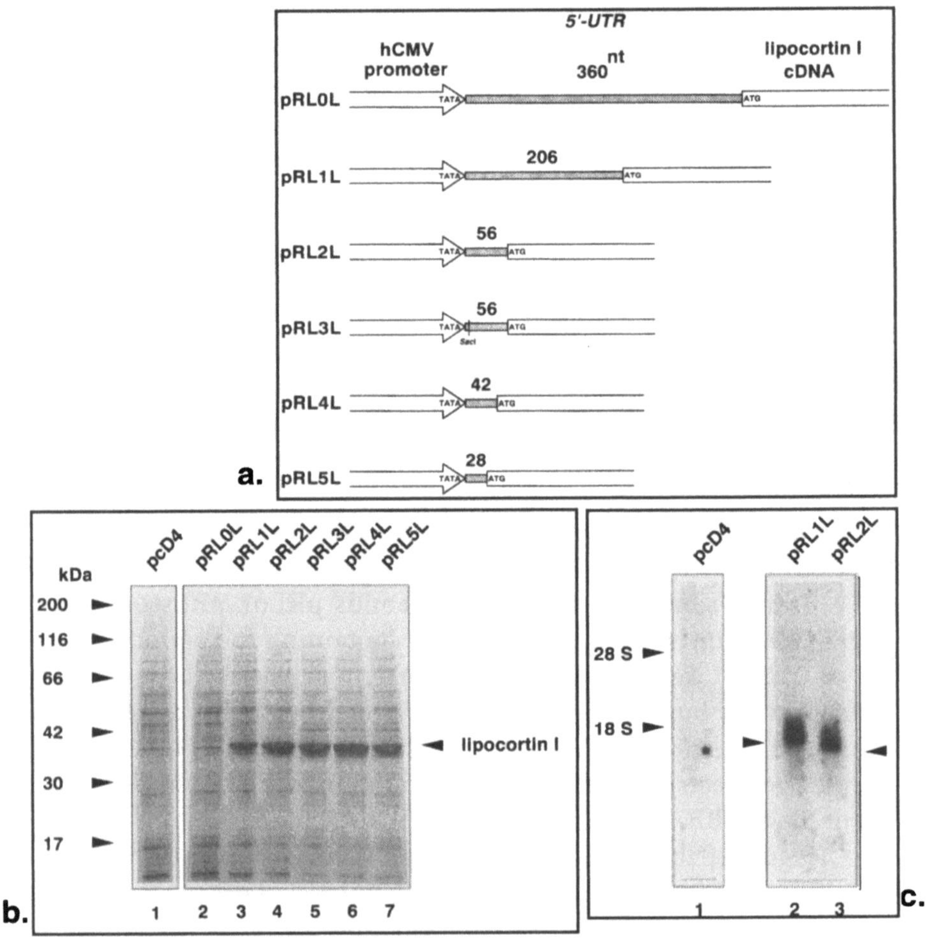

Fig. 4.2a. Structure of the *5'-UTR* (shadowed box) of plasmids pRL0L-pRL5L, the recombinant vectors for the lipocortin I expression.
b. Comparison of the production levels of lipocortin I in the cell-free lysates of recombinants carrying various lengths of *5'-UTR*. Total soluble protein extracts were analyzed by SDS-PAGE and stained with CBB.
c. Comparison of the amount of mRNA expressed by lipocortin I cDNA in the transformants carrying pRL1L or pRL2L. 18S: 18S-rRNA, 28S: 28S-rRNA.

PVDF membrane for Northern blotting. The mRNA level was probed with the [γ-^{32}P]ATP-labeled *Eco*RI-*Hin*dIII fragment of human lipocortin I cDNA from pRL2L. As seen by comparing lanes 2 and 3 in Fig. 4.2c, the amount of each transcript was almost same. The difference in mobility was due to their length of *5'-UTR*.

It appeared that the translation efficiency was affected by differences in the length of *5'-UTR* of each mRNA. To interpret these data, it is important to keep in mind that the structure of the optimized *5'-UTR* could have at least two effects. First, the 5'-terminus of the mRNA may be stabilized against nuclease. In addition, the structure itself might activate the translation initiation complex. These hypotheses, however, remain to be investigated.

The final nucleotide sequence of the *5'-UTR* region is as follows (indicated by underlining): 5'-TATATAAGCAGATTTCTCTTTAGTTCTTTGCAAGAAGGTAGAGATAAAGACACTTTTTCAAACATG-3'. To obtain high translation activity, this sequence would be applied to other types of vectors for the junction of the promoter region and initiation codon.

Characterization of Recombinant Lipocortin I Produced in *S. pombe*

To demonstrate the effectiveness of these two improvements in expression vector (the copy number-controlled vector system and the optimization of the *5'-UTR*), recombinant human lipocortin I was produced, purified, and characterized as follows.

The Level of Lipocortin I Depends on the G418 Concentration in the Medium

The *S. pombe leu1-32 h*$^-$ strain was transformed by the lithium acetate method together with the recombinant vector pRL2L (Fig. 4.3a) and the *Pst*I-linearized transducing vector, pAL7. After the Leu-positive clones were screened by the MMA plate, the G418-resistant clones were screened using a YEA plate in the presence of 25 μg/ml of G418.

Recombinant *S. pombe* carrying pRL2L or pcD4 were cultivated in the presence of 0, 25, 50, 100, 200 or 400 μg/ml of G418 in YPD medium. The level of lipocortin I was dependent on the concentration of G418, as shown in Fig. 4.3b. The amount of the main band, a 37 kDa protein, increased gradually in the G418 concentration range of 0-100 μg/ml, and was saturated at a concentration of 100-400 μg/ml. Densitometric analysis showed that the production level of

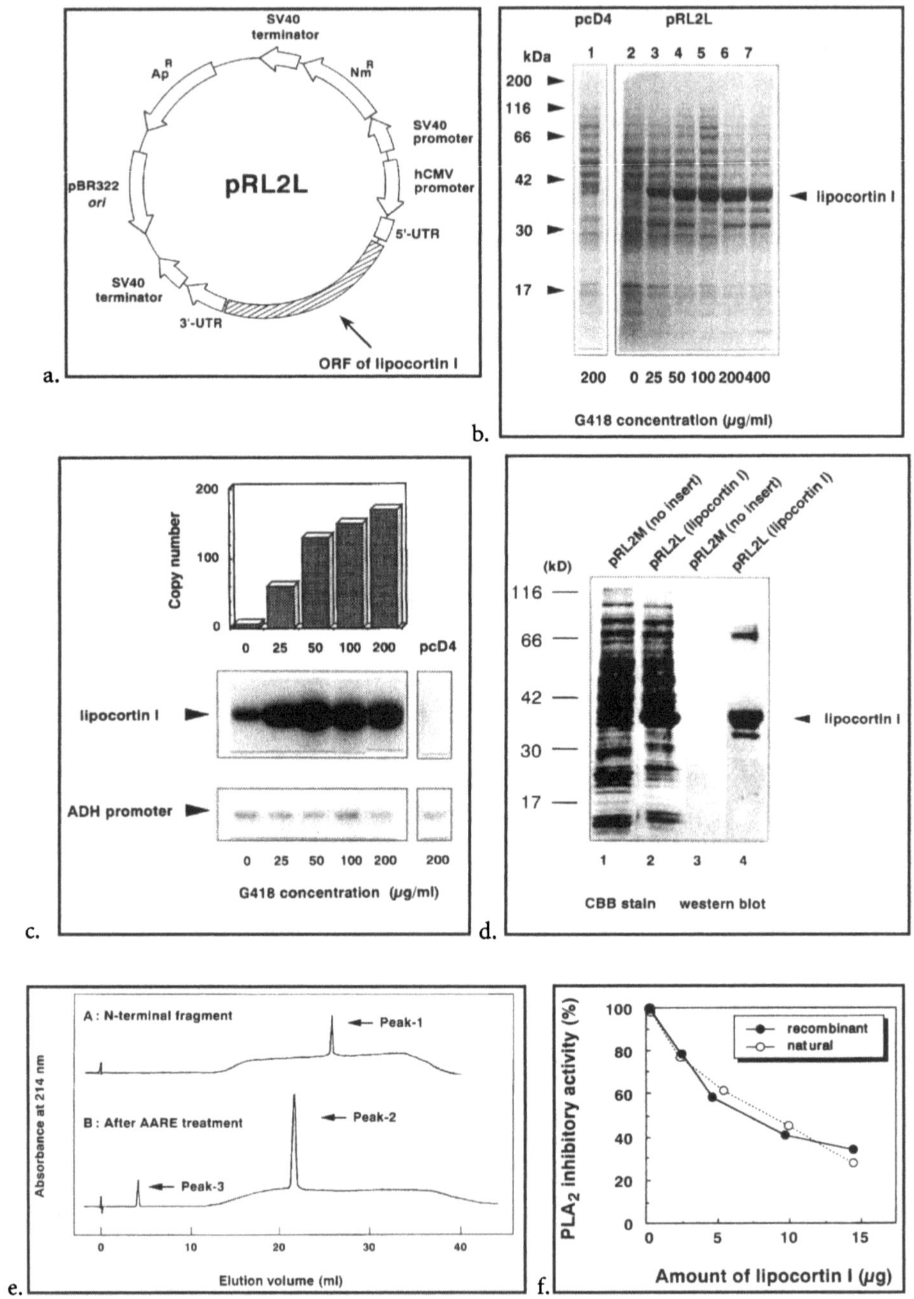
SV40 terminator
ApR
NmR
SV40 promoter
hCMV promoter
pBR322 ori
pRL2L
5'-UTR
SV40 terminator
3'-UTR
ORF of lipocortin I
a.
pcD4
pRL2L
kDa
1 2 3 4 5 6 7
200
116
66
42
30
17
lipocortin I
200 0 25 50 100 200 400
G418 concentration (μg/ml)
b.
Copy number
200
100
0
0 25 50 100 200 pcD4
lipocortin I
ADH promoter
0 25 50 100 200 200
G418 concentration (μg/ml)
c.
pRL2M (no insert)
pRL2L (lipocortin I)
pRL2M (no insert)
pRL2L (lipocortin I)
(kD)
116
66
42
30
17
lipocortin I
1 2 3 4
CBB stain western blot
d.
A : N-terminal fragment
Peak-1
B : After AARE treatment
Peak-2
Peak-3
Absorbance at 214 nm
0 10 20 30 40
Elution volume (ml)
e.
PLA2 inhibitory activity (%)
100
80
60
40
20
0
recombinant
natural
0 5 10 15
Amount of lipocortin I (μg)
f.

g. 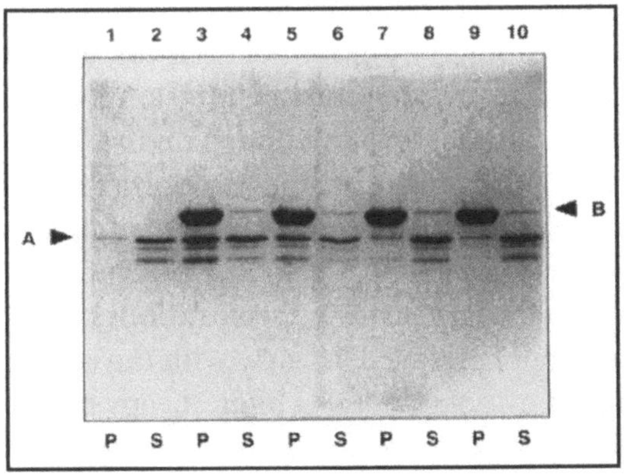

Fig. 4.3.
a. (opposite) Structure of pRL2L, the recombinant vector for lipocortin I expression.
b. Increase of the lipocortin I production level dependent on the G418 concentration in the medium. Total protein extracts were analyzed by SDS-PAGE and stained with CBB.
c. Increase of the copy number of pRL2L dependent on the concentration of G418 in the medium. Middle and lower panel show the Southern blot analysis of total DNA derived from recombinant yeast, probed with lipocortin I ORF (*Eco*RI-*Hin*dIII fragment) and *S. pombe adh* (alcohol dehydrogenase) promoter region (*Sph*I-*Eco*RI fragment), respectively. Upper panel shows the calculated copy numbers.
d. CBB stain and Western blot analysis of crude extracts of transformant carrying pRL2M (no insert control) and pRL2L (lipocortin I expression vector).
e. Determination of the N-terminus modification of recombinant lipocortin I. See the text for the detailed explanation of peaks.
f. Phospholipase A_2 inhibitory activity of recombinant and natural lipocortin I.
g. Ca^{2+}-dependent F-actin binding activity of recombinant and natural lipocortin I. SDS-PAGE analysis (CBB stain) of the pellet (P) and supernant (S) from the incubation mixture of recombinant or natural lipocortin I in buffer containing Ca^{2+} (1 mM) or EGTA (1 mM). Lanes 1 and 2: recombinant lipocortin I in the presence of Ca^{2+}. Lanes 3 and 4: recombinant lipocortin I and F-actin in the presence of Ca^{2+}. Lanes 5 and 6: natural lipocortin I and F-actin in the presence of Ca^{2+}. Lanes 7 and 8: recombinant lipocortin I and F-actin without Ca^{2+}. Lanes 9 and 10: natural lipocortin I and F-actin without Ca^{2+}. Arrowheads A and B show the position of lipocortin I and F-actin, respectively.

recombinant lipocortin I was more than 50% of the total proteins in the presence of 100-400 µg/ml of G418 (data not shown).

Figure 4.3c shows the results of a Southern blot analysis of *S. pombe* carrying pRL2L cultivated in YPD media in the absence or presence of G418. Yeast DNAs were prepared according to the method of Hoffman and Winston[16] with slight modifications. Cells, 2×10^8, were disrupted by Triton-SDS-phenol-chloroform and vortexed with glass beads (Sigma, St. Louis, MO). After RNase digestion, DNAs were digested with *Eco*RI and *Hin*dIII, resolved on 0.9% agarose gels and transferred to a PVDF membrane (Bio-Rad Laboratories, Hercules, CA) for Southern blotting. PVDF blots were probed with an

[α-^{32}P]dCTP-labeled *Eco*RI-*Hin*dIII fragment of lipocortin I cDNA from pRL2L. After autoradiography, the amount of each corresponding band was determined using the standard calibration curve of the same fragment. The copy number was calculated on the basis of the amount of the alcohol dehydrogenase promoter region (700 bp) probed by the [α-^{32}P]dCTP-labeled *Sph*I-*Eco*RI fragment of pEVP11, assuming that each cell has one copy of this region.[3] The band density increased in proportion to the increasing concentrations of G418, indicating that the copy number increased linearly with the concentration of G418 in the range of 0-100 μg/ml. The highest copy number was calculated as 170 per cell, using the *adh* promoter as a standard, in the presence of 200 μg/ml of G418; the copy number was reduced to 25 per cell in the absence of G418. As shown by the results presented in Figs. 4.3b and 4.3c, the increasing level of lipocortin I was correlated with the copy number of the recombinant vector. These results suggested that the level of heterologous protein produced was dependent on the copy number, which was controlled by the concentrations of G418 in the medium.

Western Blot Analysis of the Produced Lipocortin I

Recombinant *S. pombe* carrying pRL2L or pRL2M (a control plasmid containing multiple cloning sites instead of lipocortin I cDNA) were cultivated in YPD medium containing 200 μg/ml of G418. Cell extracts were subjected to SDS-PAGE under reducing conditions, and a 37 kDa protein specific to the cell extract of *S. pombe* carrying pRL2L was detected (Fig. 4.3d). However, when Western blots were performed using an anti-human lipocortin I polyclonal antibody, additional bands were evident. It is likely that the 68 kDa and 34 kDa proteins detected in the Western blots are the dimmer and the degradation product of lipocortin I, respectively.

The level of lipocortin I was determined by western blotting with purified placental lipocortin I as a standard, and a mouse anti-human lipocortin I monoclonal antibody and ^{125}I-labeled protein A as the first antibody and second antibody, respectively. The production level of lipocortin I was approximately 50% of the total soluble protein as measured by gamma counting (data not shown). From this estimation, in 10 grams (wet weight) of recombinant *S. pombe* cells from a 1 liter culture, the amount of expressed lipocortin I was thought to be 350 mg, because the total soluble protein was estimated to be about 700 mg.

Purification of Lipocortin I Produced in S. pombe

Recombinant lipocortin I was purified from *S. pombe* using an ion-exchange column followed by affinity chromatography. Crude extract from *S. pombe* passed through a DEAE-cellulose column and was adsorbed to a mouse anti-human lipocortin I monoclonal antibody (L2) coupled to an agarose affinity column. The adsorbed fraction was eluted with 2 mM EGTA in 50 mM Tris-HCl (pH 7.5) containing 0.15 M NaCl. The eluted protein was analyzed by SDS-PAGE, and the quantitative densitometry of CBB-stained gels indicated a purity of >95% (data not shown).

The amino acid composition analysis of the purified recombinant lipocortin I gave results essentially identical to that deduced from the cDNA of lipocortin I.[6]

Analysis of the N-Terminal Acetylation

The N-terminal Ala of human lipocortin I from human placenta has been reported to be acetylated.[17] To determine whether this modification was present in the *S. pombe* synthesized protein, recombinant lipocortin I digested with *Staphylococcus aureus* V8 protease was subjected to reverse-phase high performance liquid chromatography (HPLC) on an ODS C18 column. The N-terminal fragment was analyzed using HPLC and further treated with acylamino-acid-releasing enzyme (AARE). As shown in Fig. 4.3.e, the main peak (peak-2) now eluted faster than the intact N-terminal fragment (peak-1), and a new peak (peak-3) was also detected. The amino acid compositions of each fragment (peak-2 and peak-3) were analyzed, and it was found that peak-3 was composed of only Ala, while peak-2 contained four amino acids (Met, Val, Ser and Glu). The amino acid sequence of the latter major peak was determined to be Met-Val-Ser-Glu. The smaller peak (peak-3) was shown to consist of acetyl-Ala and not formyl-Ala (data not shown). This result shows that the posttranslational N-terminal modification of recombinant lipocortin I synthesized in *S. pombe* is identical to that of the natural human placental lipocortin I.

Biochemical Characterizations

The phospholipase A_2 (PLA_2)-inhibitory activity of recombinant and natural lipocortin I was examined using the purified recombinant lipocortin I (as described above). Native lipocortin I was purified from human placenta as described by Hayashi et al.[18] PLA_2 activity was monitored as the release of radioactivity into the assay supernatant from [^{3}H]oleic acid-labeled *E. coli* as described in Giga-Hama et al.[19] The purified recombinant protein inhibited the activity of bee venom PLA_2 (30 µg) in a dose-dependent manner with a half-maximal activity similar to that of natural placental lipocortin I (Fig. 4.3.f).

The purified lipocortin I can interact with F-actin in a Ca^{2+}-dependent manner in vitro. When the protein was incubated with rabbit skeletal muscle F-actin in the presence of Ca^{2+}, it coprecipitated with F-actin (Fig. 4.3.g). Under the same conditions, natural placental lipocortin I also binds to F-actin to a similar extent.[18,19] Without Ca^{2+}, neither recombinant nor natural lipocortin I coprecipitated with F-actin.

These results show that the recombinant lipocortin I produced in *S. pombe* is identical to that of natural proteins regarding post-translational modifications and biochemical activity.

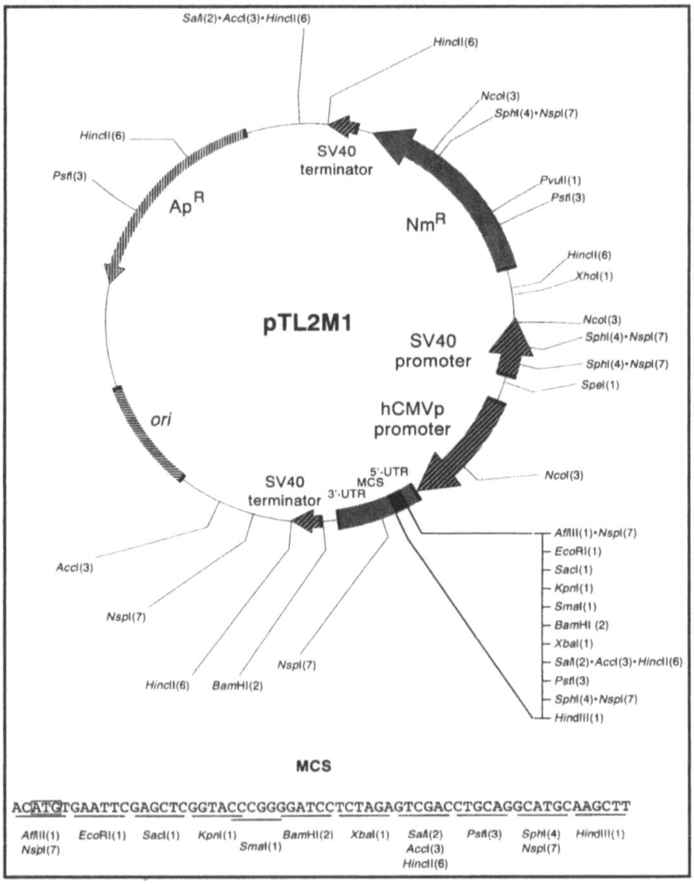

a.

Additional Expression Examples Using an ATG-Vector

To get high level expression of foreign genes it is important to maintain the length of the *5'-UTR*. To prevent the addition of amino acids to or their deletion from the N-terminus of a recombinant protein, we have constructed an ATG-vector by modifing pRL2M. Using this vector, some proteins were also produced at a highlevel without an addition of extra peptides or deletion.

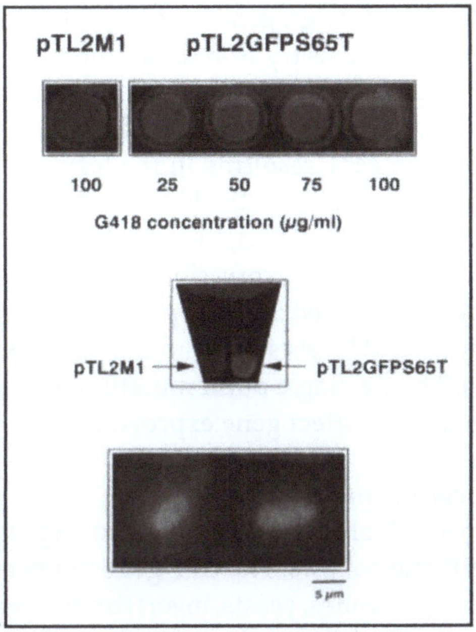

b.

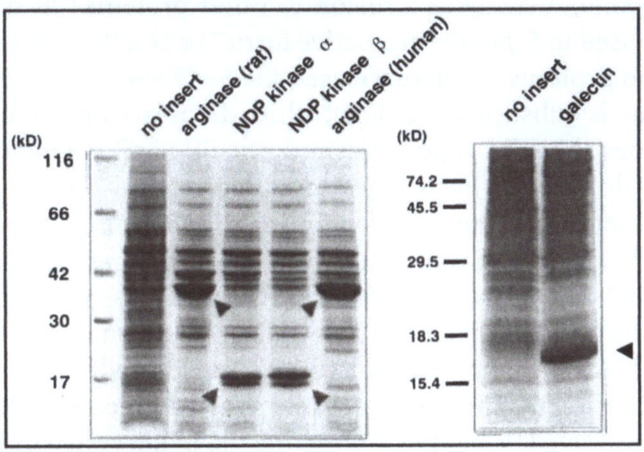

c.

Fig. 4.4a. (opposite page) Structure of the expression vector pTL2M1. The numbers in parentheses are the numbers of the restriction sites.
b. Detection of the fluorescence of GFP expressed in recombinant yeast. Transformants in a culture plate (upper panel) and harvested cells (middle panel) showed green fluorescence under long-wave UV light. Each cell also fluoresces (lower panel) under fluorescent microscopy (Olympus, Tokyo, Japan). See color figure, page 179.
c. SDS-PAGE (CBB stain) analysis of the production levels of rat and human arginase, rat NDP-kinase α and β subunits, and nematode galectin.

Construction of the Expression Vector pTL2M1

The plasmid pRL2M (section 4.4.2) was modified to give an ATG-vector as described below. A point mutation was created in the -2 position of the *5'-UTR* of lipocortin I cDNA to make an *Afl*III/*Nsp*I site for ligation with the 5'-terminus of the foreign gene. The resulting expression vector pTL2M1 (Fig. 4.4.a) also contains multiple cloning sites for ligation with 3'-terminus. The initiation ATG of pTL2M1 is part of the *Afl*III/*Nsp*I recognition site (AC<u>ATG</u>T). The initiation codon of the foreign gene must be inserted in-frame using this restriction site with "ATG-enzymes" (*Nco*I, *Bsp*HI, *Afl*III, *Sph*I; see Table 4.1).

When the initiation ATG of the foreign gene is not a part of the ATG-enzyme recognition site, it is necessary to mutate some bases around ATG to make a "ATG-enzyme" site. For example, when the open reading frame (ORF) of the foreign gene starts at ATG<u>G</u>, CC must be introduced just before the ATG, resulting in the formation of *Nco*I site (<u>CC</u>ATGG) as shown in Table 4.1. In the case of ATG<u>A</u>, ATG<u>T</u> or ATG<u>C</u>, the recognition site of *Bsp*HI, *Afl*III or *Sph*I, respectively must be introduced to construct the expression vector.

When lipocortin I cDNA was inserted into the pTL2M1 vector, recombinant protein was produced at levels similar to that obtained with pRL2L (data not shown). Thus, a single point mutation of A to C before the ATG start codon did not affect gene expression.

Expression of Green Fluorescent Protein

To utilize the ATG-vector pTL2M1, we attempted to express green fluorescent protein (GFP)[20] as follows. The GFP gene has been expressed in various organisms (microbes, yeasts, invertebrates, vertebrates and plants), usually as a fusion to other proteins.[21] It has also been expressed in *S. pombe* in a native form[22] or as a fluorescent tag of the fusion proteins.[23,24] Here we used the S65T variant of GFP (GFPS65T, Ser65 is substituted to Thr), which fluoresces more intensely than does the wild-type.[25] Plasmid phGFP-S65T (Clontech Laboratories, Palo Alto, CA) was digested with *Nco*I and *Xba*I, and an approximately 700 bp fragment containing the ORF of GFPS65T was purified. This fragment was cloned into pTL2M1 digested with *Afl*III and *Hin*dIII. *S. pombe* (*leu1-32 h^-* strain) was transformed with this resultant plasmid (pTL2GFPS65T), and the transformants were screened by MMA minimum media and YEL liquid medium containing 10μg/ml of G418. The transformants showed green fluorescence using long-wave UV light in culture media, in harvested cells

Table 4.1 Set of the compatible cohesive ends

Insert	Mutated insert (ATG-enzyme)	Vector (Compatible enzyme)	Result
nnATG**G**	c^cATGG (*Nco*I)	A^CATGT (*Afl*III)	acATG**G**
nnATG**A**	t^cATGA (*Bsp*HI)	A^CATGT (*Afl*III)	acATG**A**
nnATG**T**	a^cATGT (*Afl*III)	A^CATGT (*Afl*III)	acATG**T**
nnATG**C**	gcATG^C (*Sph*I)	ACATG^T (*Nsp*I)	acATG**C**

and in each cell (Fig. 4.4.b). After cultivation in YPD containing 200 µg/ml of G418, cells were harvested and broken by glass beads. Cell-free lysate was analyzed by SDS-PAGE. The expression level was 10% of total proteins (data not shown).

Expression Examples of Several Other Proteins

Using pTL2M1, several other foreign genes were expressed. Five recombinant vectors were constructed by a method similar to that described above. Figure 4.4.c illustrates the production levels of liver arginase, both from human[26] (from Japanese Cancer Research Resources Bank (JCRB), Tokyo, Japan) and rat[27] (a kindly gift from Drs. Mori and Kawamoto, Kumamoto University Medical School, Kumamoto, Japan), and of rat NDP-kinase α[28] and β[29] subunits (the kind gift of Dr. Kimura, Tokyo Metropolitan Institute of Gerontology, Tokyo, Japan), and nematode galectin[30] (kindly given by Dr. Hirabayashi, Teikyo University, Kanagawa, Japan). Densitometric scanning of the gels showed that the arginases, two subunits of NDP-kinase were produced at levels corresponding to 30-50% of the soluble proteins. To construct recombinant vectors, the optimal *5'-UTR* sequence of pTL2M1 and the intact N-terminus region of cDNA were maintained. This strategy resulted in high production levels of these recombinant proteins.

The specific activity of each recombinant protein was compared with the corresponding natural protein. In every case, the specific activities were close to those of the natural proteins purified from the organisms. These results indicate that, using *S. pombe*, we can obtain recombinant proteins which can be used for biochemical purposes.

Conclusions

We have constructed a novel expression vector which shows the high-level expression of various types of heterologous proteins. To achieve this high-level expression, two improvements were mainly used: the first was to use the copy number-controlled system for the maintenance of recombinant plasmid vectors in yeast cells. Using the weak *neo* expression cassette, the copy number depended on the concentration of G418 in the medium. This is the cause of the selective cell growth, due to which low-copy-number cells cannot grow in relatively high concentrations of G418. The second was the improvement of translation efficiency. By using the best length of the *5'-UTR*, the translation activity was maximized.

Using this expression vector in *S. pombe*, we obtained human lipocortin I, human and rat arginase, rat NDP kinase, nematode galectin and jellyfish GFP at the level of 10-50% of total cellular proteins. All expressed proteins were biochemically active and correctly modified in their structure. In the case of lipocortin I, the recombinant protein showed PLA_2 inhibitory and actin binding activity at the same levels as those of the natural protein purified from human placenta. Its N-terminal methionine residue was released followed by acetylation. This structure is the same as that of the natural protein. For arginase, NDP kinase and galectin, the biochemical activity was also the same as that of the corresponding natural proteins (data not shown). The N-terminal methionine of galectin was not released as is the natural protein (unpublished result). The recombinant yeast cells expressing GFP showed green fluorescence. Thus, the structure of the chromophore is correctly formed.

Future Perspectives

Recombinant protein production systems using living cells or organisms are currently the best means of producing proteins; other methods such as chemical synthesis and cell-free synthesis have significant limitations. Several useful host-vector systems have already been used. However, the perfect system has not yet been developed. For the selection of the most appropriate expression system, there are two frequently asked questions:

1) Can the expression system produce target proteins quickly, easily and cheaply?
2) Can the system produce proteins that are biologically active in the same manner and to the same degree as the natural proteins?

The use of yeast expression systems has provided some answers to these questions. Well-used yeasts such as *S. cerevisiae*, *Pichia pastoris* and *Hansenula polymorpha* are safe organisms; some have been used in the food industry or for the production of pharmaceuticals. The manipulation of yeasts is not complicated and their cultivation is not expensive or difficult. In the case of *S. cerevisiae*, the entire sequence of the chromosomal DNA has been analyzed.[31]

S. pombe is also a harmless eukaryote categorized as Class-I by the American *Type Culture* Collection,[32] i.e., it "consists of agents of no recognized hazard under ordinary conditions". Fission yeast cells grow quickly and are easy to manipulate in the laboratory. The DNA sequencing of three *S. pombe* chromosomes is being performed by the cooperative efforts of three groups, one each in Europe (The Sanger Centre, UK; Cold Spring Harbor Laboratory), the United States and Japan (National Institute of Radiological Sciences). The total sequence of the mitochondrial genome of *S. pombe* has been analyzed and added to the nucleotide sequence databases under accession no. X54421. Other information is also available from some references[15,33-35] and internet homepages (see the links in http://flosun.salk.edu/~forsburg/lab.html).

For the production of proteins in a biochemically active form, adequate posttranslational modifications and the correct tertiary and quaternary structure is necessary. With the use of *S. pombe*, the N-terminal of recombinant lipocortin I is acetylated. The fluorescence of recombinant GFP indicates that its three-dimensional structure was correctly formed. By using secretion signals as described in chapter 11, modifications in the secretory pathway are also possible. Based on the finding of Yasumori (chapter 7) that recombinant cytochrome P450 contains a heme molecule, this expression system is thought to be useful for the production of active hemoproteins or metalloproteins. The drug-metabolizing activity of P450 was increased by the expression of P450 reductase with cytochrome P450 molecules in the same cell (unpublished result).

S. pombe is an attractive host in the field of molecular and cellular biology. A heterologous protein production system in *S. pombe* has been achieved using various plasmid vectors, as summarized in chapter 2. We have successfully expressed several cDNAs using a pTL2M1 vector. There is evidence that *S. pombe* splices a mammalian gene using the same sites.[4] It may be possible to directly express functionally unknown genomic DNAs that have putative introns. From all of the standpoints examined, we conclude that the *S. pombe* host-vector system is one of the most powerful systems now available for heterologous protein production.

Acknowledgments

This work was performed with Drs. Hirofumi Okada, Yuko Giga-Hama and Hiromichi Kumagai, Asahi Glass Company. We wish to thank Dr. M. Koji Owada, Kyoto Pharmaceutical University, for the determination of the production level of lipocortin I. We are also grateful to Dr. Hiroto Okayama, Tokyo University, for his valuable discussion and his gift of the plasmid pcD4.

References

1. Losson R, Lacroute F. Plasmids carrying the yeast OMP decarboxylase structural and regulatory genes: transcription regulation in a foreign environment. Cell 1983; 32:371-377.
2. Russell PR, Hall BD. The primary structure of the alcohol dehydrogenase gene from the fission yeast *Schizosaccharomyces pombe*. J Biol Chem 1983; 258:143-149.
3. Russell P. Gene cloning and expression in fission yeast. In: Nasim A, Young P, Johnson BF, eds. Molecular Biology of the Fission Yeast. San Diego: Academic Press, Inc, 1989; 243-271.
4. Käufer NF, Simanis V, Nurse P. Fission yeast *Schizosaccharomyces pombe* correctly excises a mammalian RNA transcript intervening sequence. Nature 1985; 318:78-80.
5. Toyama R, Okayama H. Human chorionic gonadotropin α and human cytomegalovirus promoters are extremely active in the fission yeast *Schizosaccharomyces pombe*. FEBS Lett 1990; 268:217-221.
6. Giga-Hama Y, Tohda H, Okada H et al. High-level expression of human lipocortin I in the fission yeast *Schizosaccharomyces pombe* using a novel expression vector. Bio/Technology 1994; 12:400-404.
7. Tohda H, Okada H, Giga-Hama Y et al. A copy-number-controlled expression vector for the fission yeast *Schizosaccharomyces pombe*. Gene 1994; 150:275-280.
8. Wallner BP, Mattaliano RJ, Hession C et al. Cloning and expression of human lipocortin, a phospholipase A2 inhibitor with potential anti-inflammatory activity. Nature 1986; 320:77-81.
9. Prasher DC, Eckenrode VK, Ward WW et al. Primary structure of the *Aequorea victoria* green-fluorescent protein. Gene 1992; 111:229-233.

10. Okayama H, Berg P. A cDNA cloning vector that permits expression of cDNA inserts in mammalian cells. Mol Cel Biol 1983; 3:280-289.
11. Chen C, Okayama H. High-efficiency transformation of mammalian cells by plasmid DNA. Mol Cell Biol 1987; 7:2745-2752.
12. Boshart M, Weber F, Jahn G et al. A very strong enhancer is located upstream of an immediate early gene of human cytomegalovirus. Cell 1985; 41:521-530.
13. Okazaki K, Okazaki N, Kume K et al. High-frequency transformation method and library transducing vectors for cloning mammalian cDNAs by *trans*-complementation of *Schizosaccharomyces pombe*. Nucleic Acids Res 1990; 18:6485-6489.
14. Heyer WD, Sipiczki M, Kohli J. Replicating plasmids in *Schizosaccharomyces pombe*: Improvement of symmetric segregation by a new genetic element. Mol Cel Biol 1986; 6:80-89.
15. Moreno S, Klar A, Nurse P. Molecular genetic analysis of fission yeast *Schizosaccharomyces pombe*. In: Guthrie C, Fink GR eds. Methods in enzymology. Guide to yeast genetics and molecular biology. San Diego: Academic Press Inc, 1991; 194:795-823.
16. Hoffman CS, Winston F. A ten-minute DNA preparation from yeast efficiently releases autonomous plasmids for transformation of *Escherichia coli*. Gene 1987; 57:267-272.
17. Biemann K, Scoble HA. Characterization by tandem mass spectrometry of structural modifications in proteins. Science 1987; 237:992-998.
18. Hayashi H, Owada MK, Sonobe S et al. Characterization of two distinct Ca^{2+}-dependent phospholipid-binding proteins of 68-kDa isolated from human placenta. J Biol Chem 1989; 264:17222-17230.
19. Giga-Hama Y, Yohda M, Kumagai H. Intracellular distribution of a 32-kDa calcium-dependent phospholipid-binding protein from human placenta. Cell Struc Func 1989; 14:587-595.
20. Heim R, Prasher DC, Tsien RY. Wavelength mutations and posttranslational autoxidation of green fluorescent protein. Proc Natl Acad Sci USA 1994; 91:12501-12504.
21. Cubitt AB, Heim R, Adams SR et al. Understanding, improving and using green fluorescent proteins. Trends Biochem Sci 1995; 20: 448-455.
22. Atkins D, Izant JG. Expression and analysis of the green fluorescent protein gene in the fission yeast *Schizosaccharomyces pombe*. Curr Genet 1995; 28:585-588.
23. Nabeshima K, Kurooka H, Takeuchi M et al. p93dis, which is required for sister chromatid separation, is a novel microtubule and spindle pole body-associating protein phosphorylated at the Cdc2 target sites. Genes Devel 1995; 9:1572-1585.
24. Sawin KE, Nurse P. Identification of fission yeast nuclear markers using random polypeptide fusions with green fluorescent protein. Proc Natl Acad Sci USA 1996; 94:15146-15151.
25. Heim R, Cubitt AB, Tsien RY. Improved green fluorescence. Nature 1995; 373:663-664.

26. Kidd JR, Dizikes GJ, Grody WW et al. A PnuII RFLP for the human liver arginase (ARG1) gene. Nucleic Acids Res 1986; 14:9544.
27. Kawamoto S, Amaya Y, Murakami K et al. Complete nucleotide sequence of cDNA and deduced amino acid sequence of rat liver arginase. J Biol Chem 1987; 262:6280-6283.
28. Ishikawa N, Shimada N, Munakata Y et al. Isolation and characterization of a gene encoding rat nucleoside diphosphate kinase. J Biol Chem 1992; 267:14366-14372.
29. Shimada N, Ishikawa N, Munakata Y et al. A second form (β isoform) of nucleoside diphosphate kinase from rat. J Biol Chem 1993; 268:2583-2589.
30. Hirabayashi J, Ubukata T, Kasai K. Purification and molecular characterization of a novel 16-kDa galectin from the nematode *Caenorhabditis elegans*. J Biol Chem 1996; 271:2497-2505.
31. Goffeau A, Barrell BG, Bussey H et al. Life with 6000 genes. Science 274;546-567.
32. Jong S-C, Edwards MJ eds. ATCC YEASTS Nineteenth edition. Rockville: American *Type Culture* Collection, 1995.
33. Nasim A, Young P, Johnson BF eds. Molecular biology of the fission yeast. San Diego: Academic Press Inc, 1989.
34. Alfa C, Fantes P, Hyams J et al. Experiments with fission yeast: A laboratory course manual. Plainview: Cold Spring Harbor Laboratory Press, 1993.
35. Yamamoto M. Fission yeast. In: Barr PJ, Brake AJ, Valenzuela P eds. Yeast genetic engineering. Stoneham: Butterworth Publishers, 1989:53-64.

Part III

Examples of Heterologous Protein Production in Fission Yeast

CHAPTER 5

Overexpression of Proteins with Seven Transmembrane Spanning Segments

Volker Hildebrandt and Dieter Oesterhelt

Introduction

Polytopic cell surface proteins with seven transmembrane spanning segments (7TMS) form a large and functionally diverse superfamily.[1] Up to now more than 400 genes of this superfamily have been cloned and the total number is expected to exceed 2000. For many members of this superfamily, it has been shown that the N-terminus is on the extracellular side of the membrane and the C-terminus is on the intracellular side. The proposed secondary structure is exemplified for bacteriorhodopsin (BR) in Figure 5.1. The transmembrane segments are believed to be α-helices (23-26 amino acids long), which are oriented roughly perpendicular to the membrane. BR is a light-driven proton pump of the Archeon *Halobacterium salinarium.*[2,3] It is currently the only protein of the 7TMS superfamily whose structure has been elucidated at 3.5 Å resolution in the direction parallel and about 10 Å perpendicular to the membrane plane.[4] This makes BR a topographic model for other proteins with 7TMS, even for those of Eukarya.[5-9] The resolution of bovine rhodopsin is about 9 Å parallel and approximately 47 Å perpendicular to the plane of the membrane.[10,11] The primary sequence information on potential members of the superfamily is rapidly increasing and as a consequence, the suspected sequence identity between the proteins is decreasing.[12,13] The superfamily can be roughly divided

Foreign Gene Expression in Fission Yeast Schizosaccharomyces pombe, edited by Yuko Giga-Hama and Hiromichi Kumagai.

into three subgroups (Table 5.1). Members of each subgroup have a higher degree of sequence identity in certain local regions which can be used for gene cloning and functional investigations. To subfamily I belong retinal containing ion pumps from Archaea (all-trans retinal) as well as the retinal containing receptors (visual rhodopsins:11-cis retinal) of higher organisms.[14-16] Interestingly, no transporter or receptor with 7TMS has been identified in bacteria such as *Escherichia coli.*

In all visual rhodopsins, the antagonist (11-cis retinal) is bound as a protonated Schiff's base to the ε-amino group of a lysine residue within the seventh TMS (helix G). These retinal containing receptors represent a special group of chromoproteins, where the bound antagonist switches to the agonist (all-trans retinal) upon absorbance of light. The structure is related to the agonists and antagonists of the biogenic amine receptors, with a cationic ammonium group at one end of the molecule and a catechol ring (or a similar structure) at the other.[17-19] All receptors that bind positively charged monoamines such as acetylcholine, histamine and adrenaline have an aspartate residue positioned close to the center of the third transmembrane domain (helix C), which is also the case for BR (D85) (Fig. 5.2) and invertebrate rhodopsins. In mammalian rhodopsins, a glutamate residue (E113) in a similar position acts as a counter ion to the positively charged Schiff'base. In the subgroup of the retinal containing receptors, the antagonist is nestling in a pocket of the receptor (Fig. 5.3). Helix D interacts with the ring structure of the antagonist. In bacterial rhodopsins and in visual pigments of invertebrates, such as cephalopoda or diptera, the agonist remains bound to the protein after the photochemical reaction and is then thermoreversibly converted to the initial state or needs a second photon for reconversion.

Table 5.1. Subfamilies of cell surface proteins with 7TMS

Subfamily	I.	II.	III.
Amino acid residues	230-500 aa	700-800 aa	400-600 aa
Extracellular domain	Small N-Terminal region	Large N-Terminal domain	Large N-Terminal domain
Size of ligands	Binding of small ligands in a membrane pocket	Binding of large ligands (28-45 kDa)	Binding of intermediate size ligands (4-10 kDa)

In vertebrates, however, the conformational change leads to the hydrolysis of the Schiff's base and the agonist (all-*trans* retinal) is released.[20-22] Interestingly, in bacterial and eukaryotic rhodopsins, the conformational changes are combined with the movement of protons.[23]

Receptors for large hormones, such as the follicle stimulating hormone, contain a long extra-cellular N-terminus which is considered to be important for ligand binding (subfamily II).[24,25] The receptors of subfamily III share little homology with each other and with the receptors of the other subfamilies except the 7TMS topology.

In the following paragraphs aspects of the heterologous expression of such proteins will be discussed with respect to the different abilities of budding (*Saccharomyces cerevisiae*) and fission yeast (*Schizosaccharomyces pombe*).

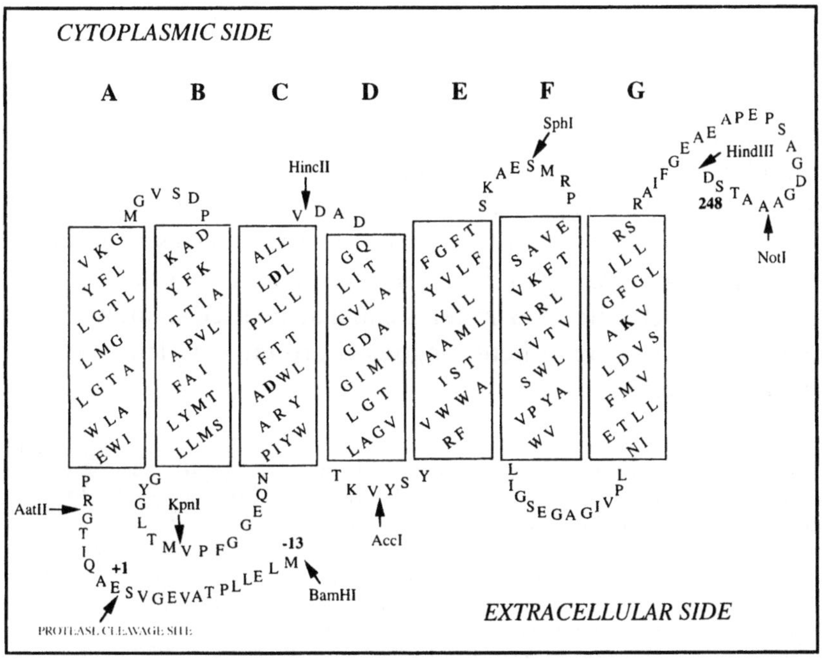

Fig. 5.1. Secondary structure model of the bacterio-opsin precursor with 262 amino acid residues. The precursor molecule contains 13 additional amino acid residues at the N-terminal end (-13 to -1) and one additional aspartate at the C-terminus (D249). The mature protein consists of 248 aa (+1 to 248). The seven helices labeled A-to-G are delineated by rectangles.[83] Single restriction sites for endonucleases in the surface exposed loops are indicated. They can be used for insertional mutagenesis studies by using bacterio-opsin as a membrane anchor for other proteins. *Bam*HI and *Hind*III restriction sites are located outside of the coding region and were used for cloning of the gene into different vectors.

G protein Coupled Receptors

Most 7TMS cell surface proteins function as sensory components of G protein coupled signal transduction cascades. These receptors bind a signaling molecule (agonist) on the extracellular side of the membrane, which leads to a conformational change of the receptor and subsequent activation of transducer molecules (heterotrimeric G proteins: αβγ) on the intracellular side.[26-28] It is generally believed that 50-60 % of all clinically relevant drugs exert their actions via G protein coupled receptors. Lanthanide ions, which inhibit the signal transduction by peptide receptors of the CCK hormone family are also inhibitors of the proton pump BR.[29] Nevertheless, G protein coupled receptors have not yet been discovered in Archaea, but lower Eukarya like fungi do have peptide receptors like animals and plants.[30-33] The principal steps of the signal transduction cascades in the two yeasts *Saccharomyces cerevisiae* and *Schizosaccharomyces pombe* (Fig. 5.4) for mating type factors are different in several aspects.[34-36] One example is the divergent evolution of the two ascomycetes to different receptor-transducer interactions. In fission yeast, the α-subunit of the G protein transmits the signal to the effector, while the βγ-complex of the heterotrimer performs this task in budding yeast.[37] In both systems, however, the ultimate result is regulation of gene expression for the purpose of controlling the mating response.[38-40] Quantification of the transduced signal is possible by fusing mating type factor induced genes with that of β-galactosidase. Such detection systems can also be used to monitor the functional expression of heterologous G protein coupled receptors in *S. cerevisiae*.[41-43]

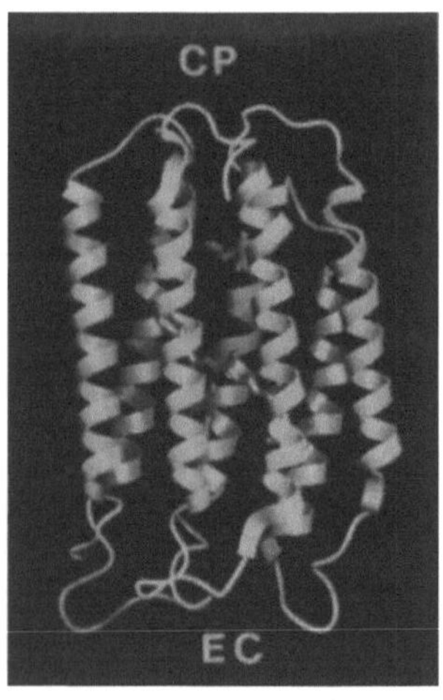

Fig. 5.2. The perpendicular view to the membrane normal shows the Cα trace of the polypeptide chain (membrane-spanning helices and loop regions) of bacteriorhodopsin. The chromophore all-trans retinal (purple) is covalently bound to K216 in helix G. The proton donor for the light induced proton translocation is D96 in the cytoplasmic channel (CP). The proton acceptor is D85 in the extracellular channel (EC). Both amino acid residues (red) are located in helix C. See color figure, page 180.

A Purple Protein as a Visible Marker

A comparison of heterologous protein expression by different vectors is based in many cases on the quantification of expressed soluble enzymes like β-galactosidase (LacZ-gene), β-glucuronidase (GUS-gene), chloramphenicol acetyltransferase (*cat*-gene) or aminoglyco-side phosphotransferase (*neo*-gene).[44-48] The reported promoter strengths are often contradictory because they can be influenced by (a) the different reporter systems, (b) the copy number, stability and segregation of the vector, (c) the repression of the reporter mRNA by an anti-sense RNA, (d) culture conditions and (e) carbon sources which regulate the rate of respiration and the rate of cell division. For the expression levels of the reporter molecule, four inherent properties of the system must be considered:

1. Half-life of the mRNA: Increasing promoter strength is occasionally correlated with decreasing half-life of the mRNA. A very stable mRNA is transcribed from the bacterio-opsin-gene (bop-gene), making it an excellent model substrate.[49-52]
2. Codon usage: The bop-gene is transcribed and translated at high levels in fission yeast without any changes in the codons while an intensive codon change is necessary for sufficient expression of the bacterio-opsin (BO) sequence in *E. coli*.[53-56]
3. Half-life of the translated protein: Expression of heterologous membrane proteins can be toxic to the host cells; in such cases the protein half-life is reduced to a few minutes. In contrast to the heterologous bacterial system, expression of membrane

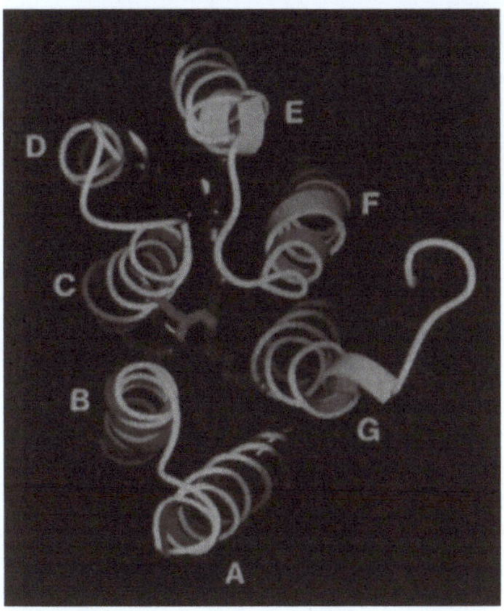

Fig. 5.3. View at the cytoplasmic surface onto bacteriorhodopsin. Helices E-G and A have stronger tilt angles. Retinal (purple) is nestling in a membrane pocket. D96 and D85 are marked in red in helix C. See color figure, page 180.

proteins in fission yeast is generally not toxic and constitutive promoters can be applied.[57]

4. Functional activity of the protein: The biological activity of membrane proteins depends on proper folding of the polypeptide and insertion into a lipid bilayer. This process can be enhanced by fusing genes with targeting sequences. Using this technique for BO in *E. coli*, (more than ten different N-terminal sequences have been tested) only a few molecules have been capable of binding retinal.[58] However, the gene transfer from the Archeon *H. salinarium* to the Eukarya *S. pombe* leads to a fully functional gene product.[59,60] In *S. pombe*, the in vivo regeneration of BO to BR is easily achieved by addition of all-trans retinal to the culture medium. Table 5.2 summarizes the differences of BO expression in *E. coli* and *S. pombe*.

The colored protein BR is a biological photochrome with a very high extinction coefficient and therefore very easy to visualize, even in the μM range.[61,62] BR consists of 248 amino acid residues and is 100 amino acid residues shorter than bovine rhodopsin, but both proteins share the same MW of 20 kDa for their membrane embed-

Fig. 5.4. Signal transduction by pheromones in haploid fission yeast cells after nutritional starvation.

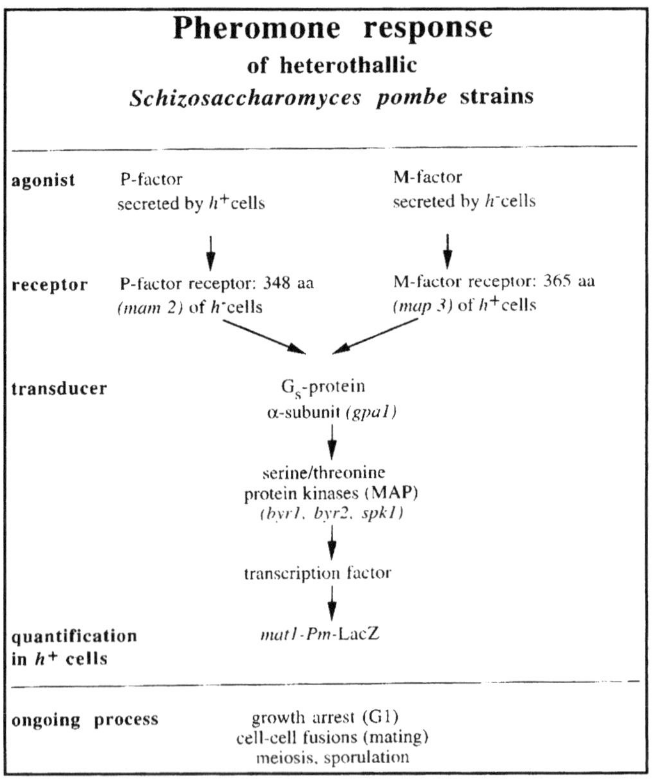

ded part.[63,64] BR has neither a N-terminal glycosylation site nor any cysteines for disulfide bridge formation and is not palmitoylated like bovine rhodopsin. The absence of these post-translational modifications makes it easier to obtain sharp protein bands on gels (SDS-PAGE) and provides the opportunity to study BR as a model for the expression of 7TMS proteins.[65] To learn more about the functional heterologous expression of integral membrane proteins, this retinal containing protein is most useful since membrane integration and functional folding can be followed by spectroscopic methods.

Tuning the Expression Level

Four strategies have been used to tune the expression level of the purple marker protein BR in fission yeast. The first two strategies summarize unpublished results of Hildebrandt and Büldt.

Integration of the Heterologous Gene

Stable integration of an expression cassette into the genome of the host is able to provide a constant expression level, which is not influenced by changing copy numbers of episomal plasmids. Genomic gene integration is therefore a useful genetic tool for increasing the expression level by using different media compositions or by modifying transcription and translation initiation regions. YIp5 is a well known integrating plasmid (5.5 kb) for *S. cerevisiae* which contains the selectable yeast URA3-gene (for a review see ref. 66). The YIp5 plasmid can also be used for integration into *S. pombe* since the URA3-gene can complement the *ura 1-D4* mutation of *S. pombe* to uracil prototrophy. YIp5 lacks sequences that allow autonomous replication of the plasmid in budding yeast, so that transformation of *S. cerevisiae* occurs by integration of the plasmid into the genome

Table 5.2. Comparison of two expression systems for a 7TMS protein

Expression of bacterio-opsin	*E. coli*	*S. pombe*
Expression is harmful	yes	no
Degradation of the protein	in min	no
Constitutive expression	no	yes
Processing of the precursor	no	yes
In vivo regeneration of the pigment	no	yes
Functional in heterologous membranes	no	yes

by recombination. In contrast, YIp5 can replicate in both *S. pombe* and in the yeast *Schwanniomyces occidentalis* at high copy numbers (A. Strasser, personal communication), but transformation efficiency of the circular vector is very low. The linearized YIp5 has an improved transformation efficiency and can integrate into the genome.

For the generation of an integrative expression vector with a strong promoter the constitutive alcohol dehydrogenase (*adh*) promoter of fission yeast was cloned into the *Sph*I and the *Hind*III site of the pBR322 backbone of YIp5.[67] The resulting integration expression vector is named pIEV and contains the polylinker of pUC18 downstream of the promoter. *Bam*HI and *Hind*III sites in the polylinker of pIEV were used for cloning the bop-gene in the appropriate orientation. The resulting vector pIBOp, which has no extra terminator, was linearized and used for transformation of fission yeast. After selection of the transformed cells, all-*trans* retinal was added to the culture medium. In vivo regeneration of BR led to purple colored yeast cells. Southern blot analysis revealed that all mRNAs transcribed from the bop-gene occurred in the same length and amount when isolated from different clones. This indicated that integration always took place at the same locus and termination of the mRNAs always occurred at the same site as well. Moreover, this integration was a stable recombination event, as evident from the fact that cells grown for more than one hundred generations in complete medium still expressed BR at the same level.

With stable integration of the expression cassette containing the *adh*-promoter and the bop-gene, the influence of the glucose concentration in the medium on the strength of the promoter could be investigated. While three alcohol dehydrogenase enzymes exist in baker's yeast (ADH1, ADH2, ADH3), there is only one alcohol dehydrogenase (*adh*) in fission yeast. It is known that some alcohol dehydrogenases can be induced with glucose. Therefore increasing concentrations of glucose were added to the media, varying from 0.5 to 10%. With increasing amounts of glucose, the purple color of the cells became more intensive, reaching a maximum at 7% glu-

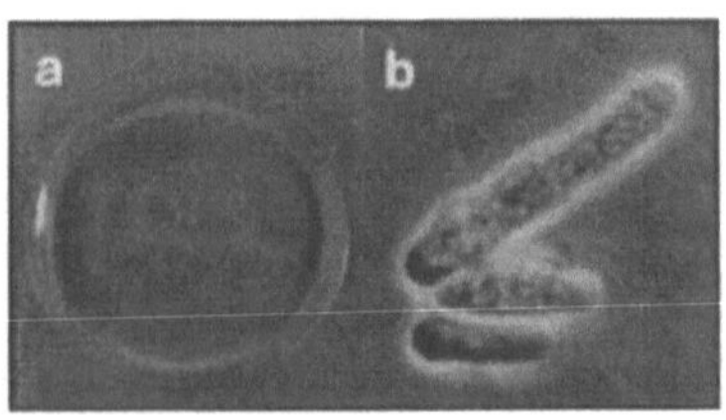

Fig. 5.5.
a. Wet cell pellet of fission yeast cells expressing bacteriorhodopsin.
b. *S. pombe* cells grown in EMM without thiamine. Two cells which overexpress bacteriorhodopsin have proliferated intracellular membranes and show an increase in size. One cell has a normal shape. See color figure, page 180.

cose. No previous reports on the influence of glucose on the *adh*-promoter have addressed the question of whether this effect is based on changed osmolarity or actually reflects specific induction. Therefore experiments were performed to test if the heterologous cauliflower mosaic virus promoter (CaMV 35S) or the homologous promoter of the thiamine-repressible gene (*nmt1*-promoter) would elevate expression levels in 7% glucose, but neither of them gave such results. Interestingly, the *fbp1*-promoter is regulated in the opposite way to that of the *adh*-promoter of *S. pombe*. Expression of the *fbp1*-gene, encoding the fructose-1,6-bisphosphatase, is inhibited by 8% glucose and derepressed at low glucose concentrations. Intermediate levels of expression can be achieved with different carbon sources.[68]

Constitutive Expression from a Stable Plasmid

High copy numbers of plasmids increase gene doses and therefore enhance expression of the desired product. Accordingly an expression cassette on a high copy plasmid is expected to give higher protein levels than the same cassette on a low copy plasmid. One of the first plasmids which was known to replicate with high copy numbers in fission yeast was pFL20, which contains the *stb*-element in conjunction with an *autonomously replicating sequence (ars)*. The *stb*-element promotes the mitotic stability of the vector in fission yeast.[69,70] The plasmid pFL20 possesses the *S. cerevisiae* URA3-gene for selection in yeast and the amp^R-gene for selection in *E. coli*.

An expression cassette useful for different hosts was isolated from the cauliflower mosaic virus. The CaMV 35S-promoter is active in many plant species, including tobacco, petunia, corn and carrot. In fission yeast, the promoter strength is comparable with the strong homologous *adh*-promoter. Transcription in *S. pombe* is initiated at a unique site identical to that used in plant tissues. The CaMV 35S-promoter had been inserted into yeast shuttle vectors like pDB262, which carries the *S. cerevisiae* 2 μm DNA but neither the *S. pombe ars* nor the *stb*-element.[71] As a consequence, pDB262 derivatives are not very stable during mitotic segregation and occur only at low copy numbers in the *S. pombe* cell. For more reproducible and stable expression levels, the CaMV 35S-promoter followed by a linker sequence and the 35S-terminator was cloned into the *Bam*HI site of pFL20. The resulting plasmid was named pFL35S and in an analogous plasmid the bop-gene was ligated to generate the vector pFLBOp. *S. pombe* transformed with pFLBOp gave higher expression levels than pIBOp, which is expected from the higher copy number of the pFL20-derivative.

Optimized Growth Rates for Unstable Vectors

Vectors based on the *S. cerevisiae* 2 µm DNA (Table 5.3) have low copy numbers in *S. pombe* cells. Due to asymmetric segregation between daughter cells they are very unstable. Even under selective conditions less than half of the cells may actually contain the vector. This plasmid loss is not easy to follow under all culture conditions and growth rates. The plasmids can exist in a partially polymeric form, making their recovery into *E. coli* very difficult. When using BR as a colored marker, however, the expression can be monitored conveniently (Fig. 5.5.a).

After cloning the bop-gene downstream of the *adh*-promoter into pEVP11, the expression of BR derived from the resulting vector pEVBOp was investigated again by in vivo regeneration of the pigment with retinal.[72] About 20 different clones were analyzed, showing variation on the level of expression, in spite of the fact that all cultures had the same growth rates. The culture with the highest purple color intensity was selected and maintained in logarithmic growth phase by appropriate dilution with fresh media. Using these

Table 5.3. Genetic differences of yeasts

Evolutionarily separated ascomycetes	***S. cerevisiae***	***S. pombe***
Regulation of cell cycle like mammals	–	+
Number of chromosomes	16	3
Cells prefer to be	diploid	haploid
Structure of small nuclear U2 RNA organization resembles those of multicellular organisms	–	+
2m plasmids occur in	high copy	low copy
Function of eukaryotic promoters:	–	+
- HIV 1	–	+
- SV 40	–	+
- CaMV 35S	–	+
Distance between TATA-box to start of transcripts	35-120 nt	25-45 nt
RNA 3' end formation related to mammalian signals	–	+
5' splicing site of the mRNA related to mammalian sites	–	+
Exons	less	many
Conjugation with *E. coli* possible	+	–
Cell fusion with mouse cells possible	–	+
Homologue to mammalian p53	–	+

intensive culturing conditions, plasmid loss can be reduced to a minimum and cells frozen in glycerol are stored at -80°C (until needed again). Highest expression levels of transformed *S. pombe* cells can be obtained with pEVP11-derivatives in YNBA minimal medium when maintained in the logarithmic growth phase. YNBA contains 0.67% yeast nitrogen base without amino acids, 0.2% asparagine, 0.2% sodium-glutamate and 7% glucose. If the cells of a culture are synchronized after repetitive heat shocks, the appearance of bacteriorhodopsin by purple color is visible in the shaker flasks.[73]

Induced Expression from a High Copy Plasmid

Induction of gene expression is important when the overexpression eventually kills the host cells. For the thiamine-repressible *nmt1*-promoter, Maundrell and coworkers employed a family of episomal (pREPs) and integrative vectors (pRIPs) with two different marker genes.[74] A disadvantage of these vectors is that the *Nde*I restriction site determines in frame fusions with the ATG-start codon of the following insert. Due to the *Nde*I-site, all these fusion vectors have an inefficient translation initiation region. To address this problem, the ATG of the polylinker of pREP1 was destroyed and the orientation of the unique restriction sites within the polylinker changed to *Bam*HI, *Sal*I, *Sma*I. In the resulting derivative (pNEU), with the LEU2-marker gene of *S. cerevisiae*, the bop-gene (pNB) was inserted and the expression level compared with previous results:

1) Based on the homologous *ars* sequence, the transformation efficiency is very high and all transformed clones growing on selection plates express BR at the same level.

Table 5.4. Bacteriorhodopsin (BR) in fission yeast

Promoter/ terminator	**Shuttle vector**	**Copy number**	**Marker gene**	**New vector**	**BR level**
adh-promoter no terminator	pYIP5	single in the genome	URA 3	pIBOp	low
CaMV-prom. and -termin.	pFL35S (pFL20)	high and stable	URA 3	pFLBOp	middle
adh-promoter no terminator	pEVP11	variable	LEU 2	pEVBOp	high
nmt 1-prom. and -termin.	pNEU (pREP1)	high and stable	LEU 2	pNB	very high

adh = alcohol dehydrogenase; CaMV = cauliflower mosaic virus, *nmt1* = no message in thiamine

2) Expression of BR begins about 14 hours after induction and reaches its maximum after 16-18 hours in EMM liquid culture medium. BR expression becomes visible in the shaker flasks upon addition of retinal to the cultures. Cells should be harvested at this time. Overexpression of this and other membrane proteins (Hildebrandt, unpublished) leads to a proliferation of almost all cellular membranes and generates bigger and longer fission yeast cells (Fig. 5.5.b). At 1000-fold magnification in a light microscope, the cells appear as if they have Bubonic plague. If cells are further incubated, the color starts to disappear. The resulting cell death can be stopped by the addition of thiamine to the culture media. After several hours the cells will recover and show normal morphology.
3) The expression level driven by the *nmt*-promoter is not influenced by the glucose concentration in the medium and is therefore useful for studying the expression of mitochondrial proteins. At high glucose concentrations, yeast starts anaerobic fermentation which causes degeneration of the mitochondria.

The expression vectors used for the expression of BO in fission yeast are summarized in Table 5.4. Even with stable replicating plasmids, cultures must be maintained in the exponential growth phase to achieve overexpression of proteins.

Future Applications

For the production of recombinant proteins in eukaryotic expression systems like fungi, insect or mammalian cell cultures, the following criteria should be examined:

1) Are the expressed proteins functionally folded?
2) Are the post-translational protein modifications close to the natural protein?
3) Is fermentation of the hosts cells possible?
4) How expensive is culturing of the cells?
5) How fast can the recombinant proteins be produced? For most of these aspects fungi can be the organisms of choice. Of particular relevance to the expression of recombinant proteins is the fact that fission yeast possesses readily visible Golgi stacks, which are reminiscent of those in higher cells. *S. pombe* can galactosylate proteins and is able to take up mevalonic acid. Therefore, glycosylation and prenylation of proteins is different from *S. cerevisiae*.[75-77] However, the even more important difference between the two ascomycetes is the amount of ex-

pressed protein: it has been shown that the expression level for BO is at least 10 times higher in fission than in budding yeast.[78] In vivo regeneration of bacteriorhodopsin could never be achieved in *S. cerevisiae* but was easily obtained in *S. pombe*. An additional comparative study with a 7TMS protein on the D_{2S} dopamine receptor revealed a 3- to 5-fold higher concentration of functional receptors in *S. pombe*.[79] This phenomenon of higher expression in fission yeast is not restricted to proteins with odd numbers of TMS, since investigations on transporters with 12 TMS gave similar results.[80,81] The reliable expression vectors (pREP-derivatives), the easiness of electroporation of competent cells, the capacity of post-translational protein modifications and the low price large scale fermentation make fission yeast a very interesting host for the production of recombinant membrane proteins.[82]

Acknowledgments

We would like to thank Dr. Georg Büldt, Dr. Ronald H. Kaback, Dr. Astrid Hoffmann, George Rosenberg, Frank Polakowski, Matthias Pfeiffer, Rita Wiemeyer, Andrew Elia and Jens Riesle for support and stimulating discussions.

References

1. Gundermann T, Nürnberg B, Schultz G. Receptors and G proteins as primary components of transmembrane signal transduction. J Mol Med 1995; 73:51-63.
2. Oesterhelt D, Stoeckenius W. Rhodopsin-like protein from the purple membrane of *Halobacterium halobium*. Nature (New Biol) 1971; 233:2853-2857.
3. Heberle J, Riesle J, Thiedemann G et al. Proton migration along the membrane surface and retarded surface to bulk transfer. Nature 1994; 370:379-382.
4. Henderson R, Baldwin JM, Ceska TA et al. Model for the structure of bacterio-rhodopsin based on high-resolution electron cryo-microscopy. J Mol Biol 1990; 213:899-929.
5. Khorana HG. Adventures in light-transducing systems: Bacterial rhodopsins and mammalian rhodopsin. In: Sarma RH, Sarma MH, eds. Structural biology: the state of the art. Proc Eighth Convers. New York: RG Adenine Press, 1993:1-17.
6. Strader CD, Fong TM, Graziano MP et al. The family of G protein coupled receptors. FASEB J 1995; 9:745-754.
7. Zhang D, Weinstein H. Polarity conserved positions in transmembrane domains of G protein coupled receptors and bacteriorhodopsin. FEBS Lett 1994; 337:207-212.

8. Röper D, Krüger P, Grötzinger J et al. Models of G protein coupled receptors revised for a family-wide compliance with experimental data. A new sequence accommodation suggested for helix G. Receptors Channels 1995; 3:97-106.
9. Hoflack J, Trumpp-Kallmeyer S, Hibert M. Re-evaluation of bacteriorhodopsin as a model for G protein coupled receptors. TiPS 1994; 15:7-9.
10. Schertler GFX, Villa C, Henderson R. Projection structure of rhodopsin. Nature 1993; 362:770-772.
11. Unger VM, Schertler GFX. Low resolution structure of bovine rhodopsin determined by electron cryo-microscopy. Biophys J 1995; 68:1776-1786.
12. Baldwin JM. The probable arrangement of the helices in the G protein-coupled receptors. EMBO J 1993; 12:1693-1703.
13. Baldwin JM. Structure and function of receptors coupled to G proteins. Curr Opin Cell Biol 1994; 6:180-190.
14. Oesterhelt D, Tittor J, Bamberg E. A unifying concept for ion translocation by retinal proteins. J Bioener Biomem 1992; 24:181-191.
15. Seidel R, Scharf B, Gautel M et al. The primary structure of sensory rhodopsin II: a member of an additional retinal protein subgroup is coexpressed with its transducer, the halobacterial transducer of rhodopsin II. Proc Natl Acad Sci USA 1995; 92:3036-3040.
16. Gärtner W, Towner P. Invertebrate visual pigments. Photochem Photobiol 1995; 62:1-16.
17. Oprian DD. The ligand-binding domain of rhodopsin and other G protein-linked receptors. J Bioener Biomem 1992; 24:211-217.
18. Parini A, Moudanos CG, Pizzinat N et al. The elusive family of imidazoline binding sites. TiPS 1996; 17:13-16.
19. Reppert SM, Weaver DR. Melatonin Madness. Cell 1995; 83:1059-1062.
20. Yarfitz S, Hurley JB. Transduction mechanisms of the vertebrate and invertebrate photoreceptors. J Biol Chem 269; 20:14329-14332.
21. Zucker CS. The biology of vision in *Drosophila*. Proc Natl Acad Sci USA 1996; 93:571-575.
22. Stryer L. Vision: from photon to perception. Proc Natl Acad Sci USA 1996; 93:557-559.
23. Arnis S, Hofmann KP. Two different forms of metarhodopsin II: schiff'base deprotonation precedes proton uptake and signaling state. Proc Natl Acad Sci USA 1993; 90:7849-7853.
24. Wallin E, Heijne G. Properties of N-terminal tails in G protein coupled receptors: a statistical study. Prot Engineering 1995; 8:693-698.
25. Coughlin SR. Expanding horizons for receptors coupled to G proteins: diversity and diseases. Curr Biol 1994; 6:191-197.
26. Lambright DG, Noel JP, Hamm HE et al. The 1.8 Å crystal structure of transducin α·GDP: Structural determinants for the activation of the α subunit of a hetero-trimeric G protein. Nature 1994; 369: 621-628.
27. Clapham DE. The G protein nanomachine. Nature 1996; 379:297-299.

28. Lambright DG, Sondek J, Bohm A et al. The 2.0 Å crystal structure of a heterotrimeric G protein. Nature 1996; 379:311-319.
29. Lutz J, Weyher E, Moroder L. Metal ion binding affinities of gastrin and CCK in membrane mimetic environment. J Peptide Sci 1995; 1:360-370.
30. Nelson MA. Mating systems in ascomycetes: a romp in the sac. TIG 1996; 12:69-74.
31. Hughes DA, Yamamoto M. *Ras* and signal transduction during sexual differentiation in the fission yeast *Schizosaccharomyces pombe.* In: Kurjan J, Taylor BL, eds. Signal transduction. London: Academic Press Inc, 1993:123-146.
32. Kitamura K, Shimoda C. The *Schizosaccharomyces pombe mam2* gene encodes a putative pheromone receptor which has a significant homology with the *Saccharomyces cerevisae* Ste2 protein. EMBO J 1991; 10:3743-3751.
33. Tanaka K, Davey J, Imai Y et al. *Schizosaccharomyces pombe* $map3^{+}$ encodes the putative M-factor receptor. Mol Cell Biol 1993; 13:80-88.
34. Eriksson OE, Svedskog A, Landvik S. Molecular evidence for the evolutionary hiatus between *Saccharomyces cerevisiae* and *Schizosaccharomyces pombe.* Systema Ascomycetum 1993; 11:179-187.
35. Huysmans E, Dams E, Vandenberghe A et al. The nucleotide sequences of the 5S rRNAs of four mushrooms and their use in studying the phylogenetic position of basidiomycetes among the eukaryotes. Nucl Acids Res 1983; 11:2871-2880.
36. Labib K, Moreno S. Rum1: a CDK inhibitor regulating G1 progression in fission yeast. T Cell Biol 1996; 6:62-66.
37. Obra T, Nakafuku M, Yamamoto M et al. Isolation and characterization of a gene encoding a G protein α-subunit from *Schizosaccharomyces pombe*: involvement in mating and sporulation pathways. Proc Natl Acad Sci USA 1991; 88:5877-5881.
38. Herskowitz I. MAP kinase pathways in yeast—for mating and more. Cell 1995; 80:187-197.
39. Hartwell L. Introduction to cell cycle controls. In: Hutchison C, Glover DM, eds. Cell Cycle Control. Oxford: IRL Press 1995:1-15.
40. Yamamoto M. The molecular control mechanisms of meiosis in fission yeast. TIBS 1996; 21:1-22.
41. Fulton R, Van Ness B. Luminescent reporter gene assay for luciferase and β-galoactosidase using a liquid scintillation counter. BioTechniq 1993; 15:119-124.
42. King K, Dohlman HG, Thorner J et al. Control of yeast mating signal transduction by mammalian β_2-adrenergic receptor and G_s α-subunit. Science 1990; 250:121-123.
43. Nielsen O, Davey J, Egel R. The *ras1* function of *Schizosaccharomyces pombe* mediates pheromone-induced transcription. EMBO J 1992; 11:1391-1392.
44. Forsburg SL. Comparison of *Schizosaccharomyces pombe* expression systems. Nucl Acids Res 1993; 21:2955-2956.

45. Hirt H. A novel method for in situ screening of yeast colonies with the β-glucuronidase reporter gene. Curr Genet 1991; 20:437-439.
46. Pobjecky N, Rosenberg GH, Dinter-Gottlieb G et al. Expression of the β-glucuronidase gene under the control of the CaMV 35S promoter in *Schizosaccharomyces pombe.* Mol Gen Genet 1990; 220: 314-316.
47. Basi G, Schmid E, Maundrell K. TATA box mutations in the *Schizosaccharomyces pombe nmt1* promoter affect transcription efficiency but not the transcription start point or thiamine repressibility. Gene 1993; 123:131-136.
48. Lang-Hinrichs C, Dössereck C, Fath I et al. Use of the Tn903 neomycin resistance gene for promoter analysis in the fission yeast *Schizosaccharomyces pombe.* Curr Genet 1990; 18:511-516.
49. Stolt P, Zillig W. Structure specific ds/ss-RNase activity in the extreme halophile *Halobacterium halobium.* Nucl Acids Res 1993; 24:5595-5599.
50. Xu ZJ, Moffett DB, Peters TR et al. The role of the leader sequence coding region in expression and assembly of bacteriorhodopsin. J Biol Chem 1995; 270:24858-24863.
51. Humphery T, Birse CE, Proudfoot NJ. RNA 3' end signals of the *S. pombe ura4* gene comprise a site determining and efficiency element. EMBO J 1994; 13:2441-2451.
52. Oliveira CC, McCarthy JEG. The relationship between eukaryotic translation and mRNA stability. J Biol Chem 1995; 270:8936-8943.
53. Dunn R, McCoy J, Simsek M et al. The bacteriorhodopsin gene. Proc Natl Acad Sci USA 1981; 11:6744-6748.
54. Sverdlov ED, Zozulya SA, Zaitseva EM et al. Genetic engineering studies of bacterioopsin gene. In: Ovchinnikov YA, ed. Proc 16th FEBS congr. Part C. VNU Science Press, 1985:131-138.
55. Karnik S, Nassal M, Doi T et al. Structure-function studies on bacteriorhodopsin. II. Improved expression of the bacterio-opsin gene in *Escherichia coli.* J Biol Chem 1987; 262:9255-9263.
56. Pompejus M, Friedrich K, Teufel M et al. High-yield production of bacteriorhodopsin via expression of a synthetic gene in *Escherichia coli.* Eur J Biochem 1993; 211:27-35.
57. Hildebrandt V, Ramezani-Rad M, Swida U et al. Genetic transfer of the pigment bacteriorhodopsin in the eukaryote *Schizosaccharomyces pombe.* FEBS Lett 1989; 234:137-140.
58. Karnik S, Doi T, Molday R et al. Expression of the archaebacterial bacterio-opsin gene with and without signal sequence in *Escherichia coli:* The expressed proteins are located in the membrane but bind retinal poorly. Proc Natl Acad Sci USA 1990; 87:8955-8959.
59. Hildebrandt V, Polakowski F, Büldt G. Purple fission yeast: Overexpression and processing of the pigment bacteriorhodopsin. Photochem Photobiol 1991; 54:1009-1016.
60. Hildebrandt V, Fendler K, Heberle J et al. Bacteriorhodopsin expressed in *Schizosaccharomyces pombe* pumps protons through the plasma membrane. Proc Natl Acad Sci USA 1993; 90:3578-3582.

61. Hampp N. Heat-proof proteins. Nature 1993; 366:12.
62. Oesterhelt D, Bräuchle C, Hampp N. Bacteriorhodopsin: a biological material for information processing. Quat Rev Biophys 1991; 24:425-478.
63. Findlay JBC, Pappin DJC. The opsin family of proteins. Biochem J 1986; 238:625-642.
64. Bridge RS. Nature of the primary photochemical events in rhodopsin and bacteriorhodopsin. Biochim Biophys Acta 1990; 1016:293-327.
65. Grisshammer R, Tate CG. Overexpression of integral membrane proteins for structural studies. Quart Rev Biophys 1995; 28:315-422.
66. Brent R, Janssen. Yeast vectors. In: Ausubel FM, Brent R, Kingston RE et al. eds. Short protocols in molecular biology. New York: John Wiley & Sons 1992; 13:19-24.
67. Russel P, Hall BD. The primary structure of the alcohol dehydrogenase gene from the fission yeast *Schizosaccharomyces pombe*. J Biol Chem 1983; 258:143-149.
68. Hoffman CS, Winston F. A transcriptionally regulated expression vector for the fission yeast *Schizosaccharomyces pombe*. Gene 1989; 84:473-479.
69. Losson R, Lacroute F. Plasmids carrying the yeast OMP decarboxylase structural and regulatory genes: transcription regulation in a foreign environment. Cell 1983; 32:371-377.
70. Chevallier MR, Lacroute F. Expression of the cloned uracil permease gene of *Saccharomyces cerevisiae* in a heterologous membrane. EMBO J 1982; 1:375-377.
71. Hirt H, Kögl M, Murbacher T, Heberle-Bors E. Evolutionary conservation of transcriptional machinery between yeast and plants as shown by the efficient expression from the CaMV 35S promoter and 35S terminator. Curr Genet 1990; 17:473-479.
72. Russel P. Gene cloning and expression in fission yeast. In: Nasim A, Young P, Johnson BF, eds. Molecular biology of the fission yeast. San Diego: Academic Press Inc, 1989:244-272.
73. Mitchison JM. Synchronous cultures and age fractionation. In: Campbell I, Duffus JH, eds. Yeast, a practical approach. Oxford: IRL Press, 1988:51-64.
74. Maundrell K. Thiamine-repressible expression vectors pREP and pRIP for fission yeast. Gene 1993; 123:127-130.
75. Chappell TG, Warren G. A galactosyltransferase from the fission yeast *Schizosaccharomyces pombe*. J Cell Biol 1989; 109:2693-2702.
76. Lehle L, Tanner W. Protein glycosylation in yeast. In: Montreuil J, Vliegenthart JFG, Schachter H, eds. Glycoproteins. Amsterdam: Elsevier Science BV 1995:475-509.
77. Giannakoruros T, Armstrong J, Magee AI. Protein prenylation in *Schizosaccharomyces pombe*. FEBS Lett 1992; 297:103-106.
78. Lang-Hinrichs C, Queck I, Büldt G et al. The archaebacterial membrane protein bacterio-opsin is expressed and N-terminally processed in the yeast *Saccharomyces cerevisiae*. Mol Gen Genet 1994; 244:183-188.

79. Sander P, Grünewald S, Reiländer H et al. Expression of the human D_{2S} dopamine receptor in the yeasts *Saccharomyces cerevisiae* and *Schizosaccharomyces pombe*: a comparative study. FEBS Lett 1994; 344:41-46.
80. Weber A, Menzlaff E, Arbinger B et al. The 2-oxoglutarate/malate translocator of chloroplast envelope membranes: molecular cloning of a transporter containing a 12-helix motif and expression of the functional protein. Biochem 1995; 34:2621-2627.
81. Caspari T, Stadler R, Sauer N et al. Structure/Function relationship of the *Chlorella* Glucose/H^+ symporter. J Biol Chem 1994; 269: 3498-3502.
82. Prentice HL. High efficiency transformation of *Schizosaccharomyces pombe* by electroporation. Nucl Acids Res 1992; 20:621.
83. Grigorieff N, Ceska TA, Downing KH et al. Electron-crystallographic refinement of the structure of bacteriorhodopsin. Submitted.

Chapter 6

Targeting and Isolation of Tagged Membrane Proteins

Volker Hildebrandt

Introduction

Membrane proteins heterologously expressed in yeast are rather stable, if correctly folded and inserted into lipid bilayers. They are resistant against proteolytic attacks by soluble proteases. Nevertheless, one often experiences difficulties expressing such proteins, especially when a transmembrane spanning segment of the protein is part of a desired fusion site. This is because few techniques exist for obtaining information about putative transmembrane spanning segments, making it difficult to precisely localize extra-membrane regions safe for the insertion of tags or targeting sequences. Computer-supported models are limited in their practical application.[1] For example, predicted membrane spanning segments based on hydropathy plots very seldomly have charged amino acid residues in the hydrophobic phase. However, charged amino acid residues are absolutely necessary here for solute transport and for stabilization of the tertiary protein structure by salt bridges.[2]

The tertiary structure and biological activity of membrane proteins is frequently influenced by the lipid composition of the surrounding bilayer.[3-5] A well studied example of lipid/protein interactions is bovine rhodopsin. This G protein coupled receptor is unable to activate transducin when embedded in the plasma membrane of rod cells because the cholesterol content (28 mol %) is too high to allow the necessary conformational changes. Only in disc membranes, where the cholesterol concentration is 8 mol %, is activation

Foreign Gene Expression in Fission Yeast Schizosaccharomyces pombe, edited by Yuko Giga-Hama and Hiromichi Kumagai.

of transducin by illuminated rhodopsin possible.[6] Another protein whose lipid/protein interactions are well-known is the purple marker protein bacteriorhodopsin (BR). Unlike bovine rhodopsin, BR functions in practically any lipid environment. However, it requires archaeal lipids for the formation of purple membranes, in which BR is most stable.[7-11] In the purple membrane BR exists as a trimer, in a two dimensional lattice, while it is a less stable monomer when expressed in yeast membranes (Table 6.1).

Interestingly, the two ascomycetes *Schizosaccharomyces pombe* and *Saccharomyces cerevisiae* differ in their lipid synthesis pathways of certain lipids.[12] Fission yeast depends on inositol in the culture medium for the synthesis of phosphatidylinositol, whereas budding yeast does not require inositol (Table 6.2). In the following paragraphs different features of the two evolutionarily separated ascomyctes will be discussed in respect to protein targeting, processing and isolation.

Protein Targeting and Processing

Since *S. pombe* propagates by fission rather than budding, vesicular traffic from the Golgi can not be organized as it is in *S. cerevisiae.*[13,14] Mechanisms of protein targeting in *S. pombe* is therefore closer to those in higher Eukarya than in *S. cerevisiae.* For example, lipid modifications of proteins are necessary for the transport processes within the cell. Since prenylation of proteins is

Table 6.1. Lipid/Protein-Interactions of Bacteriorhodopsin (BR)

Type of membrane	Purple membrane of *H. salinarium*	Plasma membrane of *S. pombe*
Aggregation	BR-trimer	BR-monomer
Mobility	immobile BR	mobile BR
Type of lipids	rigid chains, highly branched, saturated phytanyl diether lipids	ester bounds to glycerolipids
Composition of lipids	61% phosphatidylglycerol-phosphate (monomethylester)	53% phosphatidylcholine, 23% phosphatidylinositol
	26% sulfated glycolipids	11% phosphatidylethanol-amine,
	10% squalenes	6% phosphatidylserine
	3% phosphatidylglycerol	7% others

different in the two ascomycetes, the location of heterologously expressed lamin is not the same.[15] Only in fission yeast is lamin targeted correctly to the nucleus (Tamar Enoch, personal communication). However, fatty-acylated Ras-proteins are localized predominantly in the plasma membrane.[16] A wide variety of eukaryotic proteins are covalently attached to glycosyl phosphatidylinositol (GPI). The primary role of the GPI is to anchor the protein to the outer leaflet of the plasma membrane.[17] The YPT/, Rab family, which is part of the Ras superfamily and is necessary for cellular transport processes, has a higher homology to the mammalian counterparts than budding yeast.[18-22] Another aspect which has been elucidated is that *S. pombe* recognizes the ER retention signal KDEL of mammalian cells (Table 6.3).[23]

Unfortunately, there are less reports on the mechanisms of protein targeting in *S. pombe* than in *S. cerevisiae.* One reason is that only a few protein transport- and protease-deficient mutants of fission yeast are available, making it difficult to experimentally follow the processing and cellular transport of membrane proteins in this organism. Therefore, the bacterio-opsin-gene (bop-gene) was transformed into different *S. cerevisiae* strains.[24] Subcellular fractionations of transformed cells showed that bacterio-opsin (BO) was present in the plasma membrane of *S. cerevisiae* and diminished in the vacuolar membrane. Membrane integration of BO occurred in a cotranslational process and BO is also accumulated in the plasma membrane of *S. pombe.*[25,26] Western blot analysis revealed (Fig. 6.1) that budding yeast was able to process the N-terminus of the BO precursor in the same way as fission yeast and halobacteria. BO

Table 6.2. Comparison of culturing conditions for yeasts

Evolutionarily separated ascomycetes	***S. cerevisiae***	***S. pombe***
Mating type induction in	rich medium	nitrogen starvation
Growth on galactose	+	–
Growth on malic acid	–	+
Uptake of mevalonic acid	–	+
Inositol required for growth	–	+
Resistance against:		
- EtOH	+	–/+
- heat	–	+
- osmotic stress	–	+

expressed in the vacuolar-protease-deficient yeast strain cl3-ABYS86 showed the precursor exclusively. This indicates that vacuolar proteases were responsible for the formation of mature BO in *S. cerevisiae*. The vacuole houses a variety of hydrolytic enzymes and is involved in degradative and nutrient recycling functions. Degradation of plasma membrane proteins takes place in the yeast vacuole after transport via vesicular intermediates (endocytosis). Despite the fact of proper N-terminal protein processing, in vivo regeneration of BR could not be accomplished in budding yeast, but was easily achieved in *S. pombe* by the addition of retinal to the cells.[27] This may be due to the 10 times lower expression level of BO in *S. cerevisiae* than in *S. pombe*. However, it may also be caused by different interactions of the membrane protein with the protein translocation apparatus.

A key element of the translocation apparatus is the signal recognition particle (SRP).[28] SRP acts as an adapter molecule which links translation with the insertion of membrane proteins. The backbone of this particle is the SRP RNA. The fact that the binding of the SRP54 protein to this RNA is essential in fission, but not in baker's yeast, indicates one of the differences in the protein translocation apparatus of the two yeasts.[29,30]

Since budding and fission yeast differ also in their respiratory metabolism of sugars, two enzymes are located in different subcel-

Table 6.3. Comparison of yeasts at the protein level

Evolutionarily separated ascomycetes	***S. cerevisiae***	***S. pombe***
Binding of heat shock factors at temperature	normal temperature	higher
SRP RNAs resemble mam. SRP RNA	–	+
SRP 54p homologues are essential	–	+
Recognition of the KDEL signal	–	+
Homologue to mam. chaperone calnexin	–/+	+
Posttranslational modifications:		
- terminal galactose	–	+
- prenylation	–	+
Actin is closely related to mam. α-actin	–	+
Members of the YPT-family are highly homologous to mammalian counterparts	–	+
Complementation of cdc25 defect by *Xenopus* protein possible	–	+

lular compartments. The ornithine carbamoyltransferase and carbamoylphosphate synthetase are in the cytosol of *S. cerevisiae*, whereas in *S. pombe* the enzymes are targeted into the matrix of mitochondria.[31]

On the other hand, signal sequences for the post-translational import of proteins into mitochondria of *S. cerevisiae* are also recognized by *S. pombe*.[32] This has been demonstrated by incorporation of BO into the inner mitochondrial membrane of fission yeast cells. N-terminal fusion of the bop-gene with the leader sequences of the Rieske Fe/S-complex III or of the cytochrome c oxidase subunit IV of *S. cerevisiae* gave functional chromoproteins, as was also the case with the homologous leader sequence of the β-subunit of the F_1F_0-ATPase of *S. pombe*. During protein import, the signal sequences were removed and mature proteins generated.[33]

Subcellular Location of the Proteins

Of specific interest is the subcellular location of heterologous membrane proteins in a new host because the location determines the assays used to test for functional activity. In situ approaches to visualize the location of the membrane protein are electron microscopy (with gold-labeled antibodies directed against the protein) or fluorescence microscopy. For fluorescence microscopy, a variety of probes can be used (Table 6.4) to identify specific cellular compartments in fission yeast (for a review see refs. 34-36). The activity of

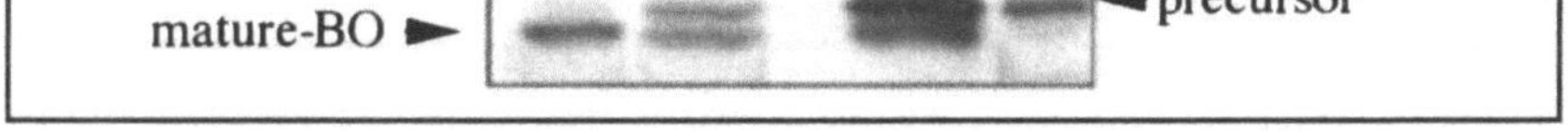

Fig. 6.1. Comparison of bacterio-opsin (BO) expression in *S. cerevisiae* and *S. pombe*. The Western blot shows a protein extract of the *S. cerevisiae* strain AH22 with mature BO in the first left lane. The protein is generated by a N-terminal truncated bop-gene, which serves as a control for completely processed BO. In the second lane the partially processed precursor molecule is also detectable. These protein species are produced by the authentic bop-gene. In the third lane, the mature and partially processed precursor proteins of *S. pombe* are visible. It is significant to note that a smaller membrane fraction was loaded for *S. pombe* than for *S. cerevisiae*, emphasizing the greater level of expression in *S. pombe*. The last lane on the right shows no mature protein. This reveals that the processing of the partially processed precursor must take place in the vacuoles of *S. cerevisiae* since the strain ABYS86 is deficient in vacuolar proteases.

transport proteins can often be monitored by changes in intracellular pH-values, as detected by pH-indicator dyes.[37] Certain dyes like BCECF (2',7'-bis(2-carboxyethyl)-5 (and -6)-carboxyfluoresceine) are taken up as a acetoxymethyl ester (BCECF/AM) into the mitochondria and subsequently cleaved in the matrix by an esterase. The fluorescence intensity of the free acids is correlated with the internal pH-values, which can be influenced by the heterologously expressed transport proteins. The size of these subcellular organelles is regulated by the culture conditions (carbon source, aeration and temperature).

Association of the heterologously expressed protein with other membranes is also influenced by the method of cell disruption. Commonly used methods for cell disruption are the French press, shearing with glass beads and osmotic lysis of protoplasts. For smaller volumes disruption with a probe sonifier is also possible, while the glass bead method is preferable when scaling up to large volumes. The degree of cell disruption is usually sufficient when a white foam causes the glass beads to adhere to the wall of the tube. For clean membrane fractions and intact organelles, osmotic lysis of protoplasts is the best technique.[38,39] When using this method, the cells should be harvested during their early exponential growth phase. To obtain almost completely digested cell walls, it is important to use an enzyme preparation from *Trichoderma harzianum* in combination with enzymes from *Arthrobacter luteus* (Table 6.5). The mere use of an enzyme preparation from *Trichoderma* is insufficient for obtaining good protoplasts because parts of the cell wall will remain intact. Protoplasts are necessary for not only subcellular fractionation of the cells but also for patch clamp measurements.[40] Protoplasts are also required for the generation of inside-out liposomes prepared from yeast plasma membrane to study transport processes.[41]

Table 6.4. Dyes for fluorescence microscopy

Target	Probe
Nuclear and mitochondrial DNA	DAPI, DNA-binding probe
Mitochondria	BCECF, pH-indicator
Vacuoles	CDCFDA, pH-indicator
Cytoplasm	SNARF, pH-indicator
F-Actin	Rhodamin-phalloidin
Cell wall	Calcofluor

After an initial purification of the subcellular compartments by differential centrifugation (with increasing centrifugal forces), an additional purification step should be performed. For plasma membrane preparations, an acid precipitation of mitochondria is helpful.[42] Additional methods are gradient centrifugations of sucrose, sucrose-ficoll or percoll, but also free flow electrophoresis, polymer phase partition and other methods can be considered.[43] The amount of a recombinant protein in a natural membrane normally does not exceed 10% of the total protein content of the target membrane. The purity of the membrane fraction should be determined by two

Table 6.5. Enzymes for the generation of protoplasts of* S. pombe *cells

Name	Source	Enzymes	Sold by
Zymolase 20T	*A. luteus*	β-1,3-glucan laminaripentaohydrolase, β-1,3-glucanase, protease, mannanase, amylase, xylanase, phosphatase	Kirin Brewery Co ICN Biomedicals
Zymolase 100T	*A. luteus*	β-1,3-glucan laminaripentaohydrolase, β-1,3-glucanase, protease, mannanase	Seikagaku Kogyo Co Wak Chemie Medical GmbH
Lyticase	*A. luteus*	β-1,3-glucanase	Sigma Chemical Co
NovoZym 234	*T. harzianum*	α-1,3-alpha-glucanase, β-1,3-glucanase, laminarinase, xylanase, chitinase,protease	Novo BioLabs
Lysing Enzymes	*T. harzianum*	cellulase, protease, chitinase	Sigma Chemical Co

Table 6.6. Marker enzymes for subcellular fractions

Cellular compartment	Enzymatic marker
plasma membrane	PM-ATPase, 5'-nucleosidase
endoplasmic reticulum	NADPH-Cyt. c-oxidoreductase, glucose-6-phosphatase
Golgi	alpha-mannosyltransferase, latent GDPase
secretory vesicles	acid phosphatase, invertase
peroxisomes	catalase
vacuoles:	
membrane:	alpha-mannosidase, V-ATPase
matrix:	carboxypeptidase Y
mitochondria:	
outer membrane:	monamine oxidase, kynurenine hydroxylase
inter membr. space:	L-lactat-dehydrogenase, adenylate kinase
inner membrane:	cytochrome c-oxidase, succinate dehydrogenase
matrix:	fumarase, isocitrate dehydrogenase

different methods and the results compared. Western blots using antibodies directed against marker proteins are appropriate for testing cross contaminations, while enzymatic assays are good for calculating the biochemical activity of marker enzymes as a measurement of organelle intactness (Table 6.6). For determination of the activity of proton or ion translocating membrane proteins, inhibitors of the naturally occurring transporters (e.g., ATPases) should be added for the suppression of endogenous transport activities.

Isolation of Membrane Proteins

Overexpression of recombinant membrane proteins in *S. pombe* will lead to massive proliferation of cellular membranes. A crude membrane fraction from such cells is a sufficient source with which to start solubilization experiments. In these initial trials, parameters such as the protein to detergent ratio, buffer composition, time, temperature, pH, salt concentration, concentration of reducing agent, etc. can be optimized for the selective solubilization of the expressed membrane protein. Special care must be taken to ensure the correct detergent is used (for a review see ref. 44). The quality of specific detergents varies from company to company. A quick check is possible by obtaining an absorbance spectrum of a detergent solution from 400 to 250 nm. A detergent suitable for a column purification procedure should have almost no absorbance at 280 nm and no absorbance in the visible wavelength range.

Selective solubilization is possible when incomplete processing leads to the production of two protein species in yeast. There is frequently a part of the proteins which is not fully post-translationally modified. For example, one protein species is glycosylated and the other not. Nonionic detergents with sugar molecules linked to the hydrophobic fatty acid chain, like n-dodecylmaltoside, solubilize glycosylated proteins better than those which are not glycosylated. These glycosylated proteins can be purified via lectin affinity chromatography. If all of the protein is needed, including the incompletely processed proteins, other affinity chromatography methods can be applied for *S. pombe* proteins. For most of these methods, special tags must be cloned into the coding sequence to enable one-step purification of recombinant proteins.

Large Tags

For the lactose permease of *E. coli*, tags could be inserted into almost every surface exposed region of the polytopic membrane protein without drastic reduction of the activity of the transport protein.[45] Epitope tagging is also possible for BO.[46] An extremely useful tool, especially for very hydrophobic proteins, is the insertion of a stretch of one hundred hydrophilic amino acid residues which contain the biotin acceptor domain from the oxaloacetate decarboxylase of *Klebsiella pneumoniae*.[47,48] These chimeric proteins are biotinylated in vivo in *S. pombe* (Hildebrandt, unpublished). The degree of biotinylation of recombinant proteins is close to 100% when the concentration of biotin in the culture medium is increased to 2 mg/l. Selective binding of the chimeras to immobilized monomeric avidin is then possible. Pure proteins are finally eluted with free biotin. Chimeric proteins can be easily identified without specific antibodies by commercially available test kits (Amersham Life Sci, Inc). Other large tags like LacZ or gluthathione-S-transferase (GST-fusions) can also help to solubilize recombinant membrane proteins.[49] A disadvantage for structural studies, however, may be less ordered structures of the flexible domains.

Small Tags

The smallest tag which can be eluted with free biotin is the Strep-tag carrying the deca-peptide sequence SAWRHPEFGG (Biometra Ltd).

A tag which is one amino acid residue smaller comprises the heamaglutinin (HA1)-epitope, which is recognized by the 12CA5 mouse IgG monoclonal antibody from Boehringer Mannheim GmbH. A nona-peptide of the sequence YPYDVPDYA will out-compete and elute proteins tagged with the epitope.[50] Unfortunately, the eluted protein is not as pure as the biotinylated protein. For more stringent binding of the tagged proteins to the resin the epitope must be cloned as a triplet.

The FLAG-tag from Kodak International Biotechnologies Inc. is a negatively charged octa-peptide sequence: DYKDDDDK, which can be used for immuno affinity purification. The peptide sequence contains the enterokinase cleavage site (DDDDK) and is designed for N-terminal fusions.[51]

The very small hexa-histidine tag has the advantage that it can be easily introduced to the C-terminus of integral membrane proteins, where only one in-frame fusion with the upstream sequence is necessary. In the rhodopsins, the additional 6xHis-tags have no influence on targeting and function of the chromoproteins.[52-54] A disadvantage is that membrane proteins are often only 70-80% pure after elution of the tagged proteins from immobilized metal affinity columns (IMAC). However, there are at least three different imino diacetic acids (IDA) and two different tetradentate chelator resins on the market (e.g., Ni-NTA), which allow investigators to optimize the purification procedure. Spin columns are not recommended for solubilized membrane proteins. Batch binding of the proteins to the resin is always advisable for hydrophobic proteins. Analytical detection of eluted proteins is possible with poly- and monoclonal antibodies directed against the 6xHis-tag.

If a tag hinders further investigations of the protein, the introduction of a protease cleavage site should be considered. High quality restriction endoproteases are expensive, but other proteases can be too nonspecific. A brief check for specificity should be done by incubating the protease with the purified protein before the new cleavage site is inserted into the coding sequence.

If the newly inserted cleavage site appears to be inaccessible to the protease for instance, because the site is sterically blocked by the tertiary structure of the protein, the target sequence should be repeated as a tandem or triplet.[55]

Future Perspectives

Fission yeast is able to express recombinant proteins from organisms of all kingdoms regardless of the number of transmembrane spanning segments (TMS). BR with 7 TMS, the Na^+/Ca^{2+} exchanger of the human heart muscle with 11 TMS, the lactose permease of *E. coli* with 12 TMS and even chimeric transport proteins with 19 TMS have been expressed at very high levels (Hildebrandt & Kaback, unpublished). The expressed protein can represent about 1-2 % of the total cell membrane proteins and up to 1 mg pure protein can be isolated from a 1 liter culture. In all comparative studies so far, the amount of expressed protein is remarkably higher in *S. pombe* than in *S. cerevisiae*.[56]

While the expressed eukaryotic proteins are normally functional and glycosylated, a fraction of the proteins can occur in a nonglycosylated form. Future investigations are necessary to obtain more homogeneous protein populations. One solution would be the coexpression of proteins which can facilitate protein transport through the ER and thereby increase the amount of functionally folded proteins.[57] Within the ER are enzymes that form disulfide bonds, such as protein disulfide isomerase (PDI) and peptidylprolyl isomerase (PPIase). The ER also contains chaperones such as the glucose-regulated proteins GRP94 and BiP/GRP78, calnexin, calreticulin and Hsp47.[58,59] For the homologous expression of rhodopsin1 of *Drosophila melanogaster*, the membrane bound chaperone NinaA is necessary for transport of the receptor from the ER to the plasma membrane. It can, therefore, be assumed that heterologous expression of membrane proteins can be significantly enhanced by coexpression with one of the enzymes or chaperones mentioned above.

Acknowledgments

I would like to thank Dr. Dieter Oesterhelt, Dr. Ronald H. Kaback, Dr. Paul B. Sigler, Dr. Christine Lang, Ingo Queck, Rita Wiemeyer, Bob Paca, Matthias Pfeiffer, Andrew Elia and Judith Shiozawa for support and stimulating discussion.

References

1. Eisenhaber F, Persson B, Argos P. Protein structure prediction: recognition of primary, secondary, and tertiary structural features from amino acid sequence. Crit Rev Biochem Mol Biol 1995; 30:1-94.
2. Frillingos S, Sahin-Toth M, Lengeler JW et al. Helix packing in the sucrose permease of *Escherichia coli*: properties of engineered charge pairs between helices VII and XI. Biochem 1995; 34:9368-9373.
3. Elferink MGL, De Witt JG, Driessen AJM et al. Energy-transduction properties of primary proton pumps reconstituted into archaeal bipolar lipid vesicles. Eur J Biochem 1993; 214:917-925.
4. Van der Rest ME, Kamminga AH, Nakano A et al. The plasma membrane of *Saccharomyces cerevisiae:* structure, function and bioenergetics. Microbiol Rev 1995; 59:304-322.
5. In't Veld G, Driessen AJM, Konings WN. Bacterial solute transport proteins in their lipid environment. FEMS Microbiol Rev 1993; 12:293-314.
6. Boesze-Battaglia K, Albert AD. Cholesterol modulation of photoreceptor function in bovine retinal rod outer segments. J Biol Chem 1990; 265:20727-20730.
7. Hildebrandt V, Fendler K, Heberle J et al. Bacteriorhodopsin expressed in *Schizosaccharomyces pombe* pumps protons through the plasma membrane. Proc Natl Acad Sci USA 1993; 90:3578-3582.
8. Kates M, Moldoveanu N, Steward LC. On the revised structure of the major phospholipid of *Halobacterium salinarium.* Biochim Biophys Acta 1993; 1169:46-53.
9. Watts A, Venien-Bryan C, Sami M et al. Lipid-protein interactions in controlled membrane protein array and crystal formation. In: Watts A, ed. Protein-Lipid Interaction. Amsterdam: Elsevier Sci Publ BV, 1993:351-370.
10. Jonas R, Koutalos Y, Ebrey TG. Purple membrane: surface charge density and the multiple effect of pH and cations. Photochem Photobiol 1990; 52:1163-1177.

11. Nikoloff DM, Henry SA. Genetic analysis of the yeast phospholipid biosynthesis. Ann Rev Genet 1991; 25:559-583.
12. Gaynor PM, Greenberg M. Regulation of CDP-diacylglycerol synthesis and utilization by inositol and choline in *Schizosaccharomyces pombe.* J Bacteriol 1992; 174:5711-5718.
13. Armstrong J. Two fingers for membrane traffic. Curr Biol 1993; 3:33-35.
14. Mertins P, Gallwitz D. A single intronless actin gene in the fission yeast *Schizosaccharomyces pombe:* nucleotide sequence and transcripts formed in homologous and heterologous yeasts. Nucl Acids Res 1987; 15:7369-7378.
15. Giannakoruros T, Armstrong J, Magee AI. Protein prenylation in *Schizosaccharomyces pombe.* FEBS Lett 1992; 297:103-106.
16. Masterson WY, Magee AI. Lipid modifications involved in protein targeting. In: Magee AI, Wileman T, eds. Protein Targeting, A Practical Approach. New York: IRL Press 1992:232-259.
17. Hurtley SM. Hot line to the secretory pathway. TIBS 1993; 18:3-6.
18. Miyake S, Yamamoto M. Identification of *ras*-related, *YPT* family genes in *Schizosaccharomyces pombe.* EMBO J 1990; 9:1417-1422.
19. Hengst L, Lehmeier T, Gallwitz D. The *ryh1* gene in the fission yeast *Schizosaccharomyces pombe* encoding a GTP-binding protein related to ras, rho, and ypt: structure, expression and identification of its human homologue. EMBO J 1990; 9:1949-1955.
20. Hughes DA, Ashworth A, Marshall CJ. Complementation of *byr1* in fission yeast by mammalian MAP kinase kinase requires co-expression of Raf kinase. Nature 1993; 364:349-352.
21. Newman CMH, Giannakouros T, Hancock JF et al. Post-translational processing of *Schizosaccharomyces pombe* YPT proteins. J Biol Chem 1992; 267:11329-11336.
22. Tamanoi F. Inhibitors of Ras farnesyltransferase. TIBS 1993; 18: 349-353.
23. Pidoux AL, Armstrong J. Analysis of the BiP gene and identification of an ER retention signal in *Schizosaccharomyces pombe.* EMBO J 1992; 11:1583-1591.
24. Lang-Hinrichs C, Queck I, Büldt G et al. The archaebacterial membrane protein bacterio-opsin is expressed and N-terminally processed in the yeast *Saccharomyces cerevisiae.* Mol Gen Genet 1994; 244:183-188.
25. Hildebrandt V, Bauer U, Dencher NA et al. Bacteriorhodopsin precursor is partially processed at the N-terminal end in heterologous in vitro and in vivo expression systems. In: Rigaud JL, ed. Structures and functions of retinal proteins. John Libbey Eurotext Ltd., 1992; 221:97-100.
26. Bauer U, Hildebrandt V, Dencher NA et al. In vitro synthesis of bacterio-opsin: Integration into microsomal membranes. Biochem Biophys Res Com 1992; 187:1480-1485.
27. Hildebrandt V, Polakowski F, Büldt G. Purple fission yeast: Overexpression and processing of the pigment bacteriorhodopsin. Photochem Photobiol 1991; 54:1009-1016.
28. Walter P, Johnson AE. Signal sequence recognition and protein targeting to the endoplasmic reticulum membrane. Annu Rev Cell Biol 1994; 10:87-119.

29. Ribes V, Dehoux P, Tollervey D. 7SL RNA from *Schizosaccharomyces pombe* is encoded by a single copy essential gene. EMBO J 1988; 7:231-237.
30. Selinger D, Brennwald P, Liao X et al. Identification of RNA sequences and structural elements required for assembly of fission yeast SRP54 protein with signal recognition particle RNA. Mol Cell Biol 1993; 13:1353-1362.
31. Van Huffel C, Dubois E, Messenguy F. Cloning and sequencing of *arg3* and *arg11* genes of *Schizosaccharomyces pombe* on a 10 kb DNA fragment. Eur J Biochem 1992; 205:33-43.
32. Lill R, Neupert W. Mechanisms of protein import across the mitochondrial outer membrane. T Cell Biol 1996; 6:56-61.
33. Hoffmann A, Hildebrandt V, Heberle J et al. Photoactive mitochondria: in vivo transfer of a light-driven proton pump into the inner mitochondrial membrane of *Schizosaccharomyces pombe.* Proc Natl Acad Sci USA 1994; 91:9367-9371.
34. Experiments with fission yeast. In: Alfa C, Fantes P, Hyams J, et al. Cold Spring Harbor Laboratory Press 1993.
35. Armstrong J, Fawell E, Pidoux A. Intracellular trafficking in fission yeast. In: Magee AI, Wileman T. eds. Protein Targeting, A Practical Approach. New York: Oxford Univ Press, 1992:87-112.
36. Clark MW. Immuno-electron microscopy. In: Johnston J, ed. Molecular Genetics of Yeast, A Practical Approach. New York: Oxford Univ Press, 1994:233-246.
37. Preston RA, Murphy RF, Jones EW. Assay of vacuolar pH in yeast and identification of acidification-defective mutants. Proc Natl Acad Sci USA 1989; 86:7027-7031.
38. Moreno S, Klar A, Nurse P. Molecular genetic analysis of fission yeast *Schizosaccharomyces pombe.* In: Guthrie C, Fink G, eds. Guide to Yeast Genetics and Molecular Biology. Meth Enzymol 194, Academic Press, Inc, 1991:795-826.
39. Kambe-Honjoh H, Yoda K, Yamasaki M. Protein transport in the permeabilized cell of *Schizosaccharomyces pombe.* Biochem Biophys Res Com 1992; 186:838-845.
40. Vacata V, Höfer M, Larsson HP et al. Ionic channels in the plasma membrane of *Schizosaccharomyces pombe*: Evidence from patch-clamp measurements. J Bioener Biomem 1992; 24:43-53.
41. Ruetz S, Raymond M, Gros P. Functional expression of P-glycoprotein encoded by the mouse *mdr3* gene in yeast cells. Proc Natl Acad Sci USA 1993; 90:11588-11592.
42. Dufour JP, Amory A, Goffeau A. Plasma membrane ATPase from the yeast *Schizosaccharomyces pombe.* In: Fleischer S, Fleischer B, eds. Biomembranes Part Q. Meth Enzymol 157, Academic Press, Inc. 1988:513-528.
43. Powell MJ, Watts FZ. Isolation of a gene encoding a mitochondrial HSP70 protein from *Schizosaccharomyces pombe.* Gene 1990; 95: 105-110.

44. Zulauf M. Detergent phenomena in membrane protein crystallization. In: Michel H, ed. Crystallization of Membrane Proteins. Boca Raton:CRC Press, Inc. 1990:54-72.
45. McKenna E, Hardy D, Kaback HR. Insertional mutagenesis of hydrophilic domains in the lactose permease of *Escherichia coli*. Proc Natl Acad Sci USA 1992; 89:11954-11958.
46. Teufel M, Pompejus M, Humbel B et al. Properties of bacteriorhodopsin derivatives constructed by insertion of an exogenous epitope into extra-membrane loops. EMBO J 1993; 12:3399-3408.
47. Consler TG, Persson BL, Jung H et al. Properties and purification of an active biotinylated lactose permease from *Escherichia coli*. Proc Natl Acad Sci USA 1993; 90:6934-6938.
48. Cronan JE. Biotinylation of proteins in vivo. J Biol Chem 1990; 265:10327-10333.
49. Leatherwood J, Lopez-Girona A, Russel P. Interaction of Cdc2 and Cdc18 with a fission yeast ORC2-like protein. Nature 1996; 379: 360-363.
50. Field J, Nikawa JI, Broek D et al. Purification of a RAS-responsive adenyl cyclase complex from *Saccharomyces cerevisiae* by use of an epitope addition method. Mol Cell Biol 1988; 8:2159-2165.
51. Su X, Prestwood AK, McGraw RA. Production of recombinant porcine tumor necrosis factor alpha in a novel *E. coli* expression system. BioTech 1992; 13:756-762.
52. Janssen JJM, Bovee-Geurts PHM, Merkx M, DeGrip WJ. Histidine tagging both allows convenient single-step purification of bovine rhodopsin and exerts ionic strength-dependent effects on its photochemistry. J Biol Chem 1995; 270:11222-11229.
53. Kamp Hansen O, Pompejus M, Fritz HJ. Two-chain bacteriorhodopsin synthesized by *Schizosaccharomyces pombe*. Biol Chem Hoppe-Seyler 1994; 375:715-719.
54. Baker EK, Colley NJ, Zucker CS. The cyclophilin homologue NinaA functions as a chaperone, forming a stable complex in vivo with its protein target rhodopsin. EMBO J 1994; 13:4886-4895.
55. Sahin-Toth M, Dunten RL, Kaback RH. Design of a membrane protein for site-specific proteolysis: properties of engineered factor Xa protease sites in the lactose permease of *Escherichia coli*. Biochem 1995, 34:1107-1112.
56. Loddenkötter B, Kammerer B, Fischer K et al. Expression of the functional mature chloroplast triose phosphate translocator in yeast internal membranes and purification of the histidine-tagged protein by a single metal-affinity chromatography step. Proc Natl Acad Sci USA 1993; 90:2155-2159.
57. Kell DB, Goodarce R, Neal MJ. Improving secretion of recombinant proteins from yeast and mammalian cells: rational or empirical design? TIBTECH 1994; 12:432-435.
58. Jannatipour M, Rokeach LA. The *Schizosaccharomyces pombe* homologue of the chaperone calnexin is essential for viability. J Biol Chem 1995; 270:4845-4853.
59. Nagata K. Hsp47: a collagen-specific molecular chaperone. TIBS 1996; 21:23-26.

CHAPTER 7

Expression of a Human Cytochrome P450 Form in *Schizosaccharomyces pombe*: Comparison with Expression in *Saccharomyces cerevisiae*

Toshio Yasumori

Introduction

Heterologous expression of drug-metabolizing enzymes using yeast has been conveniently conducted with *Saccharomyces cerevisiae*, and has been shown to yield the expressed amounts of the enzymes at the maximum of approximately 1% of the total cellular protein.[1] In particular, cytochrome P450s expressed in *S. cerevisiae* have been shown to elucidate catalytic activity and the substrate specificity among P450 forms with high similarity.[2] The efficiency of expression has not been greatly improved, although several *S. cerevisiae* strains and several kinds of the promoters were tested. Connecting the 5'-flanking sequence of the cDNA with the downstream sequence of the promoters has not been successful.[3] The fission yeast, *Schizosaccharomyces pombe*, is taxonomically and evolutionarily distant from *S. cerevisiae*, and can carry out post-transcriptional modifications such as glycosylation, acetylation, and phosphorylation, which have not been done in *S. cerevisiae*.[4] However, it is unclear whether or not these modifications alter the chemical and catalytic properties of cytochrome P450. There have been few reports on heterologous gene expression of drug-metabolizing and related enzymes from higher eukaryotes in *S. pombe*.[5,6]

Foreign Gene Expression in Fission Yeast Schizosaccharomyces pombe, edited by Yuko Giga-Hama and Hiromichi Kumagai.

A major advantage to the use of *S. pombe* is that the amount of expressed protein can be expected to be extremely higher than that expressed in *S. cerevisiae*. Furthermore, various kinds of promoters can be used in expression vectors for heterologous expression in *S. pombe*.[7] Although the promoters from *S. cerevisiae* can be used for *S. pombe*, the amount of expressed protein was rather small.[8] In the expression vector described here, a heterogeneous cDNA is placed under the control of the human cytomegalovirus (hCMV) promoter, which was previously found to be 10-fold more active than the SV40 early promoter.[9,10] As an example of the potential of this system, we describe the production of a human liver cytochrome P450 in this section. The yield of CYP2C9, a member of the CYP2C subfamily of cytochrome P450, was at levels exceeding 10% of the total cellular protein. The use of the hCMV promoter for the expression of CYP2C9 in *S. pombe* was compared with that in *S. cerevisiae* with the *GAL7* promoter.[1,11] Furthermore, functional CYP2C9 was obtained in *S. pombe*, with chemical and catalytic properties almost identical to those expressed in *S. cerevisiae* and the purified human protein. Multiple forms of CYP2C isozymes are expressed in human livers, and catalyze hydroxylations of various kinds of drugs. Several CYP2C9 forms are reported[2] and showed differences in their catalytic activities, but shared the identical amino-terminal protein sequence with each other, and these were reported to be indistinguishable by extensive chemical analysis. We previously reported that the expression of microheterogeneous CYP2C9 cDNAs in *S. cerevisiae* and determination of the catalytic activity are very useful to iden-

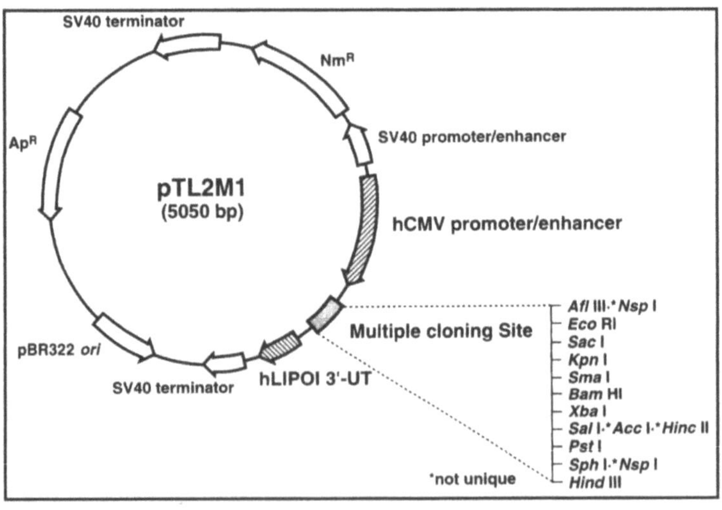

Fig. 7.1. Structure of expression vector pTL2M1.

tify each P450 with microheterogeneity.[2] However, to obtain a large amount of cytochrome P450, it was necessary to use a large-scale culture. In this section, we describe the expression of CYP2C9 cDNA in *S. pombe* using an expression vector containing the hCMV promoter (Fig. 7.1). This method has an advantage compared to using conventional *S. cerevisiae* in the yield of the expressed protein. By using this method, CYP2C9 cDNA was demonstrated to encode a major metabolic enzyme for various drugs from a single preparation in a small-scale culture.

Plasmid Constructin, Transformation and Assay

Construction of the Plasmid for Expression for CYP2C9 in S. pombe

The recombinant plasmid pTF450,[1] which directs synthesis of the CYP2C9 protein in *S. cerevisiae* cells, is shown in Figure 7.2. Starting from pTF450, the plasmid was digested with *Nco*I and *Xba*I. The digested CYP2C9 cDNA fragment was ligated with pTL2M1 (Fig. 7.1), which was digested with *Nsp*I and *Xba*I, for insertion between the hCMV promoter and 3' untranslated region of human lipocortin I, and the expression plasmid pTL2CYP2C9 was constructed.

Transformation of S. pombe *and Expression of CYP2C9 in* S. pombe

S. pombe strain [genotype: *leu1-32h*$^-$ (ATCC 38399)] was transformed with the plasmid pTL2CYP2C9, together with *Pst*I digested

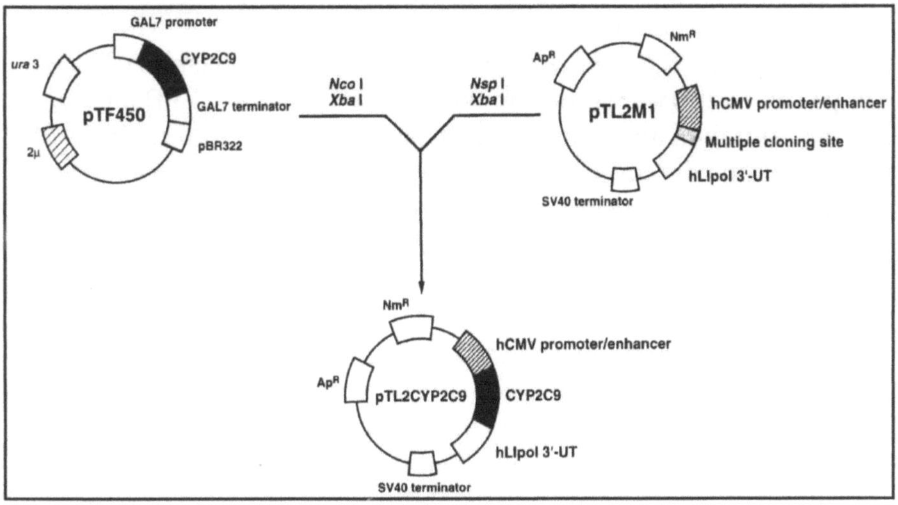

Fig. 7.2. Construction of the plasmid for expression of CYP2C9 cDNA.

transducing vector pAL7 according to Okazaki et al.[12] Transformed cells were spread on MMA minimum culture plates without leucine, incubated at 32°C for 3 days, transferred to a YEA plate containing 25 µg/ml of G418 (Gibco Laboratories), and cultured at the same temperature for 3 to 5 days. Transformed cells were grown in YPD medium (1% Bacto yeast extract (Difco, Detroit, MI, USA)/2% Bacto peptone (Difco)/2% glucose) in the presence of 100 µg/ml of G418 at 32°C until the early stationary phase. Cells were washed twice with distilled water and suspended in 50 mM Tris-chloride (pH 7.5) at 4-fold volume of cell pellet. The cells were disrupted using a sonicator on ice, and boiled at 95°C in the presence of 1% sodiumdodecylsulfate. After centrifugation, supernatant was used for immunoblot analysis.

Construction of the Plasmid for Expression of Human NADPH-Cytochrome P450 Reductase in S. pombe

Starting from the plasmid pGEMR450, the 5'-flanking region of the coding sequence of human NADPH-cytochrome P450 reductase cDNA[13] was mutated by PCR to form a unique *Nco*I site. Then the plasmid was digested with *Nco*I, *Xho*II and *Xho*II, *Eco*RI separately. The resulting two fragments of the cDNA were ligated with the expression vector pTL2M1 to form the expression plasmid pTL2hCYPOR. Transformation, expression and preparation of yeast microsomes were carried out similarly as for CYP2C9, described in the former sections.

Preparation of Yeast Microsomes from Large-Scale Culture

Preparation of microsomes of *S. pombe* was performed as described previously by the method using *S. cerevisiae* with some modifications.[1] *S. pombe leu1-32h⁻* cells carrying the plasmid pTL2M1 or pTL2CYP2C9 were grown to the stationary phase in YPD medium at 32°C and then an aliquot (30 ml) was introduced into 2.0 liters of the medium containing 100 µg/ml of G418. The cells were collected at the logarithmic phase of the growth by centrifugation, and microsomes were prepared by the method described previously with minor modifications.[1] The cells were washed twice with buffer A (10 mM Tris-chloride, pH7.5, 0.65 M d-sorbitol, 0.1 mM dithiothreitol, 0.1 mM EDTA) and resuspended to less than 0.15 g of cells/ml in buffer B (10 mM Tris-chloride, pH7.5, 2.0 M d-sorbitol, 0.1 mM dithiothreitol, 0.1 mM EDTA) containing 100 mg/ml of Zymolyase 100T. The suspension was incubated with gentle shaking for 1 hr at

30°C washed twice with buffer B, and then resuspended in buffer A containing 0.1 mM phenylmethylsulfonylfluoride, 20 µg/ml of leupeptin and 5 µg/ml of chymostatin. The resulting spheroplasts were disrupted by sonication (10W, 1 min, 5 times) and homogenized in a Dounce tissue grinder. The lysate was centrifuged at 3000 x g for 15 min to precipitate cell debris, then the supernatant was centrifuged at 10,000 x g for 20 min to recover crude mitochondrial fraction. The supernatant obtained was ultracentrifuged at 125,000 x g for 90 min to separate the microsomal fraction from the soluble fraction. The sediment recovered as the microsomal pellet was resuspended in buffer A containing 0.1 mM phenylmethylsulfonylfluoride, 20 µg/ml of leupeptin and 5 µg/ml of chymostatin.

Assay

The assay procedures for tolbutamide, phenytoin and hexobarbital and mephenytoin metabolism were described previously.[1,2] The incubation mixture consisted of 4 mg of yeast microsomal protein, 1 unit of expressed human NADPH-cytochrome P450 reductase (or purified rat NADPH-cytochrome P450 reductase), 100 mM sodium phosphate buffer (pH 7.4), 0.1 mM EDTA, an NADPH-generating system (6 mM $MgCl_2$, 0.8 mM NADP, 8 mM glucose 6-phosphate, and 1 unit/ml of glucose 6-phosphate dehydrogenase), and a substrate in a final volume of 250 µl. Substrate concentrations were 0.1 mM for tolbutamide, 0.2 mM for phenytoin, and hexobarbitals, respectively. All determinations were performed in triplicate.

Miscellaneous Methods

Sodiumdodecylsulfate(SDS)-polyacrylamide gel electrophoresis was conducted according to the method of Laemmli[14] with an acrylamide concentration of 7.5%. Immunoblot analysis was performed as described by Towbin et al,[15] using anti-CYP2C9 IgG.[16] The amount of immunoreactive protein was quantitated by densitometric readings in proportion to the densities of the purified proteins on the same blot. The content of cytochrome P450 in the microsomes was determined according to the method of Omura and Sato,[17] except for the inclusion of 0.2% Emulgen 913 in the solubilization buffer. Protein content was determined by the method of Lowry et al,[18] using bovine serum albumin as the standard.

Expression of CYP2C9 in S. pombe *and Comparison with Conventional Expression in* S. cerevisiae

The plasmid pTL2CYP2C9 or pTL2M1 was introduced into *leu1-32h*$^-$ cells. Whole-cell extracts prepared from both *leu1-32h*$^-$ [pTL2CYP2C9] and *leu1-32h*$^-$ [pTL2M1] were analyzed by immunoblotting using anti-human CYP2C antibody. As shown in Figure 7.3, a protein reacting with anti-human CYP2C antibody was observed as a major band in extracts from *leu1-32h*$^-$ [pTL2CYP2C9] which migrated at a position identical to those of authentic CYP2C9 purified from a human liver[16] and CYP2C9 expressed in *S. cerevisiae.*[1] The reduced CO-difference spectra were obtained from *leu1-32h*$^-$ [pTL2CYP2C9] and *leu1-32h*$^-$ [pTL2M1] microsomes. The microsomal fraction of *leu1-32h*$^-$ [pTL2CYP2C9] showed a prominent absorption peak at 450 nm, although no appreciable peak was detected around 450 nm in the spectrum of *leu1-32h*$^-$ [pTL2M1] microsomes (Fig. 7.4.). The results, together with the immunological data, indicate that the CYP2C9 protein represents approximately 10% of the total cellular protein as the holoenzyme state. The specific content of CYP2C9 in *leu1-32h*$^-$ [pTL2CYP2C9] microsomes ranged from 0.8 to 1.3 nmol/mg protein, which were approximately 10-fold those of *S. cerevisiae* MT8-1[pTF450] (0.11 ± 0.03 nmol/mg protein).

Catalytic Activities of CYP2C9 Expressed in S. pombe

Catalytic activities of the microsomes of *S. pombe* towards various drugs are shown in Tables 7.1 and 7.2. *leu1-32h*$^-$ [pTL2CYP2C9]

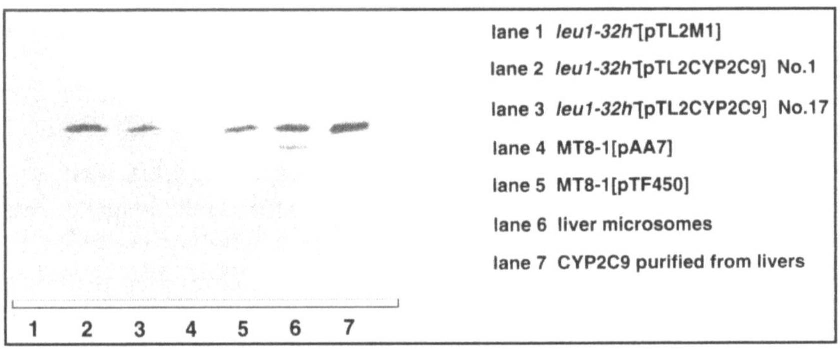

Fig. 7.3. Immunoblot analysis of CYP2C9 expressed in yeast and human liver microsomes using anti-CYP2C9 IgG. Each lane contained 2 µg protein of the whole cell extracts prepared from *leu-32h*$^-$[pTL2M1] (*lane 1*), *leu-32h*$^-$[pTL2-CYP2C9] (*lane 2, No. 1*), *leu-32h*$^-$[pTL2-CYP2C9] (*lane 3, No. 17*),and 10 µg protein from MT8-1[pAA7] (*lane 4*), MT8-1[pTF450] (*lane 5*), 2 µg proteins of liver microsomes (*lane 6*), and 0.75 pmol of purified CYP2C9 (*lane 7*) from human livers. After electrophoresis, the proteins were transferred to a nitrocellulose paper and subjected to immunoblot analysis.

microsomes catalyzed hexobarbital 3'-hydroxylations, phenytoin 4'-hydroxylation, and tolbutamide methylhydroxylation at rates similar to those shown by *S. cerevisiae* previously.[2] However, the rates of catalytic activities were approximately 10-fold faster than those in *S. cerevisiae*, because of the high-level expression of cytochrome P450 in *S. pombe*. The NADPH-cytochrome P450 reductase activity and the amount of cytochrome b_5 in *S. pombe* microsomes were very low (approximately 0.005 nmol/min/mg protein and 0.04 nmol/mg protein, respectively). The rates of reductase activity and the amounts of cytochrome b_5 were almost equivalent to those in *S. cerevisiae*. The specific activities of *leu1-32h*⁻ [pTL2CYP2C9] microsomes significantly increased with the amount of purified rat NADPH-cytochrome P450 reductase added in the incubation mixture as similar to the case of *S. cerevisiae*),[1] whereas no such enhancement of the activities was detected with control *leu1-32h*⁻ [pTL2M1]. The effect of anti-human CYP2C antibody on the activities of *S. pombe* was examined (data not shown). Anti-human CYP2C antibody strongly inhibited the activities of *leu1-32h*⁻ [pTL2CYP2C9] (>90%). However, no effect was observed on the activities of control *leu1-32h*⁻ [pTL2M1].

Discussion and Conclusions

The production of cytochrome P450 in yeast was previously performed by the use of *S. cerevisiae*.[1,19,20,21] In the present study, we have demonstrated that *S. pombe* is also a powerful host for the synthesis of cytochrome P450 in yeasts. The yield of CYP2C9 achieved, approximately 10% of the total cellular protein, appears superior to

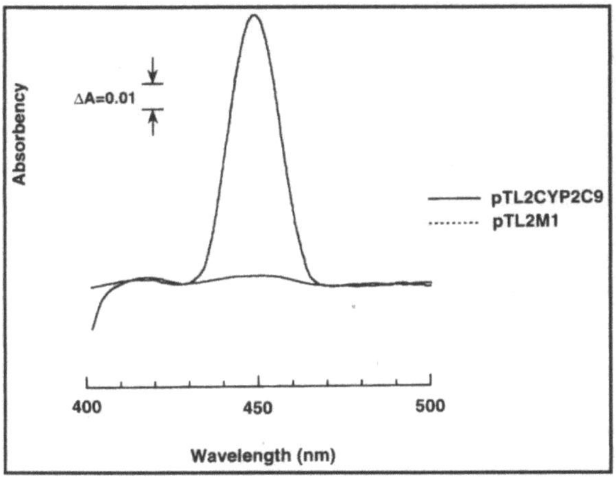

Fig. 7.4. Reduced CO-difference spectra of yeast membrane fractions. The reduced CO-difference spectra of cytochrome P450 were obtained according to the method of Omura and Sato. Spectra were recorded with 5 mg/ml of microsomal protein.

Table 7.1. (R), (S)-Hexobarbital 3'-hydroxylase and phenytoin 4'-hydroxylase activities in yeast microsomes

	(R)-Hexobarbital 3'-hydroxylation	(S)-Hexobarbital 3'-hydroxylation pmol/min/mg protein		
S. pombe *leu1-32h*$^-$[pTLCYP2C9]	256.6 ± 6.9*	169.5 ± 5.9	1.77 ± 0.03	12.98 ± 0.22
S. pombe *leu1-32h*$^-$[pTL2M1] (control)	< 0.1	< 0.1	< 0.01	< 0.01
S. cerevisiae MT8-1[pTF450]	25.1 ± 0.6	16.7 ± 0.8	0.09 ± 0.01	0.45 ± 0.01
S. cerevisiae MT8-1[pAA7] (control)	< 0.1	< 0.1	< 0.01	< 0.01

*Values represent the mean of the triplicate experiments ± SD (range). Values are the results of experiments with 1 unit of NADPH-cytochrome P450 reductase in the incubation mixture. Substrate concentrations of hexobarbitals are 0.2 mM.

Table 7.2. Tolbutamide hydroxylase and phenytoin 4'-hydroxylase activities in yeast microsomes

	Tolbutamide hydroxylation	pmol/min/mg protein	Phenytoin 4'-(R)-HPPH 4'-(S)-HPPH
S. pombe *leu1-32h*$^-$[pTLCYP2C9]	258.4 ± 23.3*	1.77 ± 0.03	12.98 ± 0.22
S. pombe *leu1-32h*$^-$[pTL2M1] (control)	< 0.1	< 0.1	< 0.1
S. cerevisiae MT8-1[pTF450]	25.3 ± 2.2	0.09 ± 0.01	0.45 ± 0.01
S. cerevisiae MT8-1[pAA7] (control)	< 0.1	< 0.01	< 0.01

*Values represent the mean of the triplicate experiments ± SD (range). Values are the results of experiments with 1 unit of NADPH-cytochrome P450 reductase in the incubation mixture. Substrate concentrations of tolbutamide and phenytoin are 0.1 and 0.2 mM, respectively. Abbrebiation: HPPH, 5-(4-hydroxyphenyl)-5-phenylhydantoin.

those expressed in *S. cerevisiae* in the previous reports. Moreover, the yield may be improved several-fold by changing the host strain of *S. pombe* or the promoter in the expression plasmid. A problem that remains to be overcome is how the catalytic activities can show rates similar to those in liver microsomes catalyzed by the identical form of P450. In order to increase the catalytic activities, we added NADPH-cytochrome P450 reductase, cytochrome b_5 and lipids in the incubation mixture. The rates of catalytic activities increased more than 3-fold, and similar enhancement of the activities was also observed by using *S. cerevisiae* in the previous study.[2] Thus, it is necessary to additionally construct the expression plasmid containing cytochrome b_5 cDNA under the control of the hCMV promoter and the enhancer. Then the catalytic activities could possibly achieve rates similar to those observed in liver microsomes. Finally, we can easily obtain a large amount of expressed cytochrome P450 in *S. pombe* from a small-scale culture. The cytochrome P450 form expressed in *S. pombe* showed chemical properties and catalytic activities almost identical to those expressed in *S. cerevisiae* and also similar to those in liver. Although several expression systems have been developed to express P450s from their cDNAs, a yeast expression system using *S. pombe* has an advantage in respect of a high yield of P450 expression, and furthermore, it can be cultured on a scale of even 1,000 liters. The large amount of expressed P450 can contribute to a large number of assays in studies of drug metabolism.

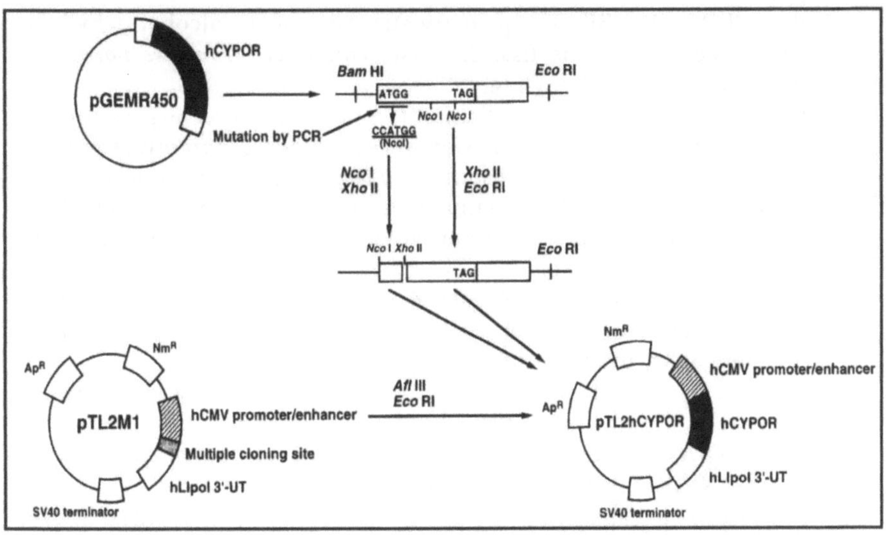

Fig. 7.5. Construction of the plasmid for expression of NADPH-cytochrome P450 reductase.

Acknowledgments

I would like to thank Drs Hiromichi Kumagai and Yuko Giga-Hama and Mr. Hideki Tohda, Asahi Glass Co., Ltd., for support and stimulaying discussion.

References

1. Yasumori T, Murayama N, Yamazoe Y et al. Expression of a human P-450IIC gene in yeast cells using galactose-inducible expression system. Mol Pharmacol 1989; 35:443-449.
2. Yasumori T, Yamazoe Y, Kato R. Cytochrome P_{450} human-2 (P-450IIC9) in mephenytoin hydroxylation polymorphism in human livers: Differences in substrate and stereoselectivities among microheterogeneous P-450IIC species expressed in yeast. J Biochem 1991; 109:711-717.
3. Kato R, Yasumori T, Yamazoe Y. Characterization of human P450IIC isozymes by using yeast expression system. Methods Enzymol 1991; 206:183-190.
4. Russell PR, Nurse P. *Schizosaccharomyces pombe* and *Saccharomyces cerevisiae*: A look at yeasts divided. Cell 1986; 45:781-782.
5. Jackson MR, Burchell B. Expression of human liver epoxide hydrolase in *Schizosaccharomyces pombe*. Biochem J 1988; 251:931-933.
6. Yamazaki S, Sato K, Suhara K et al. Importance of the proline-rich region following signal-anchor sequence in the formation of correct conformation of microsomal cytochrome P_{450}s. J Biochem 1993; 114:652-657.
7. Moreno, S, Klar A, Nurse P. Molecular genetic analysis of fission yeast *Schizosaccharomyces pombe*. Methods Enzymol 1991; 194: 795-823.
8. Russell PR, Hall BD. The primary structure of the alcohol dehydrogenase gene from the fission yeast *Schizosaccharomyces pombe*. J Biol Chem 1983; 258:143-149.
9. Giga-Hama Y, Tohda H, Okada H et al. High-level expression of human lipocortin I in the fission yeast *Schizosaccharomyces pombe* using a novel expression vector. Bio/Technology 1994;12:400-404.
10. Tohda H, Okada H, Giga-Hama Y et al. A copy-number-controlled expression vector for the fission yeast *Schizosaccharomyces pombe*. Gene 1994; 150:275-280.
11. Yasumori T, Kawano S, Nagata K et al. Nucleotide sequence of a human liver cytochrome P_{450} related to the rat male-specific form. J Biochem 1987; 102:1075-1082.
12. Okazaki K, Okazaki N, Kume K et al. High-frequency transformation method and library transducing vectors for cloning mammalian cDNAs by trans-complementation of *Schizosaccharomyces pombe*. Nucleic Acids Res 1990; 18:6485-6489.
13. Murakami H, Yabusaki Y, OhkawaH. Expression of rat NADPH-cytochrome P_{450} reductase cDNA in *Saccharomyces cerevisiae*. DNA 1986; 5:1-10.

14. Laemmli UK. Cleavage of structural proteins during the assembly of the head of bacteriophage T4. Nature 1970; 227:680-685.
15. Towbin H, Staehilin T, Gordon J. Electrophoretic transfer of proteins from polyacrylamide gels to nitrocellulose sheets: procedure and some applications. Proc Natl Acad Sci USA 1979; 76:4350-4355.
16. Kawano S, Kamataki T, Yasumori T et al. Purification of human liver cytochrome P_{450} catalyzing testosterone 6β-hydroxylation. J Biochem 1987; 102:493-501.
17. Omura T, Sato R. The carbon monooxide-binding pigment of liver microsomes. I. Evidence for its hemoprotein nature. J Biol Chem 1964; 239:2370-2378.
18. Lowry OH, Rosebrough NJ, Farr AL et al. Protein measurement with the Folin phenol reagent. J Biol Chem 1951; 193:265-275.
19. Oeda K, Sakaki T, Ohkawa H. Expression of rat liver cytochrome P_{450} MC cDNA in *Saccharomyces cerevisiae*. DNA 1985; 4:203-210.
20. Shimizu T, Sogawa K, Fujii-Kuriyama Y et al. Expression of cytochrome P_{450}d by *Saccharomyces cerevisiae*. FEBS Lett 1986; 207:217-221.
21. Fujita VS, Thiele DJ, Coon MJ. Expression of alcohol-inducible rabbit liver cytochrome P_{450} 3a(P-450IIE1) in *Saccharomyces cerevisiae* with the copper-inducible CUP1 promoter. DNA Cell Biol 1990; 9:111-118.

CHAPTER 8

Human Papillomavirus Capsid Protein-pREP in *Schizosaccharomyces pombe*: Efficient Assembly of the Viral Capsid Protein in *S. pombe* and *S. cerevisiae*

Toshiyuki Sasagawa

Introduction

Papillomaviruses infect the epithelia of a wide variety of animals and generally induce benign proliferative lesions at the site of infection. Some types of papillomavirus can induce malignant tumors. More than 70 types of human papillomavirus (HPV) have been isolated and some of these HPVs are etiological factors for various warts in the skin or the mucosal epithelium. Each HPV type has a strong host cell specificity for infection. Among the mucosal types, HPV 6 and HPV 11 are associated with benign warts, while the HPV 16[1] and the HPV 18[2] are associated with anogenital cancer and its precursor lesions.[3] Although HPV 6 or HPV 11 are generally benign in adults, they can also induce more severe lesions in children such as juvenile laryngeal papillomatosis.[4] This disease is extremely resistant to any clinical treatment. Therefore, it is important to establish human prophylactic vaccines against HPV 6 and 11, as well as the malignant

Foreign Gene Expression in Fission Yeast Schizosaccharomyces pombe, edited by Yuko Giga-Hama and Hiromichi Kumagai.

types of HPV such as HPV 16 and 18, to prevent these viral infections. The study of these viral structural proteins has progressed less than that of the oncoproteins of these viruses, because of the lack of a suitable propagation system for HPV virions in vitro.

Native virions of papillomaviruses are non-enveloped 50-60 nm spherical structures composed of 72 pentameric capsomeres in icosahedral symmetry.[5] Each capsomere is composed of five L1 molecules and form a virion shell with a T=7 symmetrical lattice. Several groups have established systems for making virus-like particles (VLP) which resemble native HPV particles but lack viral DNA. HPV 16[6] and HPV 1[7] particles have been produced in a vaccinia virus system, and HPV 11,[8] HPV 16[9] and HPV 33[10] particles in a baculovirus system. Conformational epitopes present in native HPV 11 infectious virions were also present on the baculovirus-produced HPV 11 VLP and reacted with sera obtained from individuals with condyloma acuminata.[8] These VLP may substitute for native virions of human papillomaviruses as prophylactic vaccines for humans and as antigens for ELISA predicting HPV 16 infection.[11]

Both vaccinia virus and baculovirus systems are expensive and require laborious manipulations to increase the yield of VLP. An

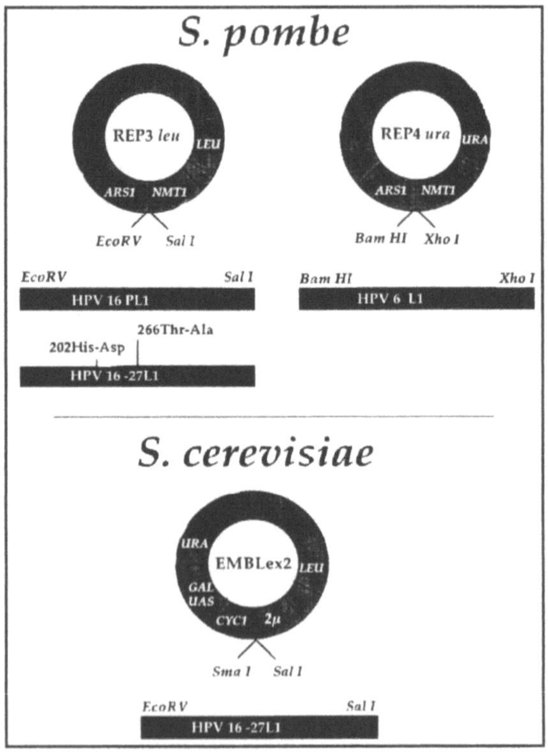

Fig. 8.1. Construction of the yeast recombinant vectors. The L1 capsid genes of HPV 16-prototype (PL1) and HPV 6 L1 were introduced in *S. pombe* vectors pREP 3 or pREP 4. The 27L1 derived from a variant HPV 16 was inserted in the REP 3 and EMBLex2 for *S. cerevisiae*. Two amino acid changes in the 27L1 from the prototype L1 are indicated.

expression system using yeast would be an attractive alternative to these higher eukaryotic systems if correctly assembled capsid proteins could be obtained. Here we describe production of HPV 16 and HPV 6 virus-like particles in the fission yeast *Schizosaccharomyces pombe* and HPV 16 VLP in the budding yeast *Saccharomyces cerevisiae* and compare the yield of the VLP in both microorganisms for practical usefulness of the VLP production.

Expression and Localization of HPV 16 Capsid Proteins in *Schizosaccharomyces pombe*

Preparation of recombinant yeast and expression of the introduced gene were carried out as follows. A prototype of HPV 6 L1 (6L1), HPV 16 L1 (PL1) and the L1 gene of HPV 16 variant (27L1)[12] were introduced into expression vectors pREP 3 and pREP 4 containing a thiamine-repressible promoter for *S. pombe* as described before.[13,14] The 27 L1 gene was introduced into the expression vector pEMBLex2 which contains the galactose-induced promoter for *S. cerevisiae* as reported by C. Baldari et al.[15] HPV 16 L1 was cloned between the *Bal*I-*Sal*I sites of pREP 3 leu and between the *Sma*I -*Sal*I sites of pEMBLex2. The HPV 6 L1 was cloned between the *Bam*HI-*Xho*I sites of pREP 4-20 leu. The details are shown in Figure 8.1. The *leu2*- and *ura3*- auxotrophic mutant *S. pombe* was transformed with the pREP-16L1 or pREP-6L1 and pREP-ura plasmids. The transformants were allowed to grow selectively in minimal essential medium without leucine and uracil supplemented with 2.5 mM thiamine. The L1-gene expression was induced by incubation at 30°C for 16 hours in the medium lacking thiamine.

The pEMBL-16L1 was introduced into the auxotrophic strain S150-2B (*ura3*-, *leu2* -, *trp1*-, *his3*-, *cir*+) of *S. cerevisiae*. The recombinant *S. cerevisiae* was selectively grown in SC-glc medium (yeast nitrogen base, nucleotides and amino acids without uracil and leucine, 2% glucose). The inserted L1 gene were induced for expression at 30°C for 16 hours in the SC-glc/gal medium (yeast nitrogen base, nucleotides and amino acids without uracil and leucine, 1% galactose, 0.3% glucose).

Native virions of human papillomaviruses accumulate in the nuclei of human epithelial cells and the capsid proteins are directed from the cytoplasm to the nucleus immediately after translation, according to Zhou et al.[16] Therefore, nuclear localization of the L1 protein is presumed to be important for viral capsid assembly or for maturation of virus particles. To investigate this in our yeast system, we performed the immunofluorescence of HPV 6 L1 and L2 proteins (Fig. 8.2) in *S. pombe*. Strongest signals of the L1 and L2 were observed in the nucleus, and relatively weak L1 and L2 signals in the cytoplasm. The localization pattern was the same in HPV 16 PL1 and 27L1 as with HPV 6L1. Nuclear localization of HPV16 L1 was confirmed by immunoblot analysis on the nuclear extract separated from

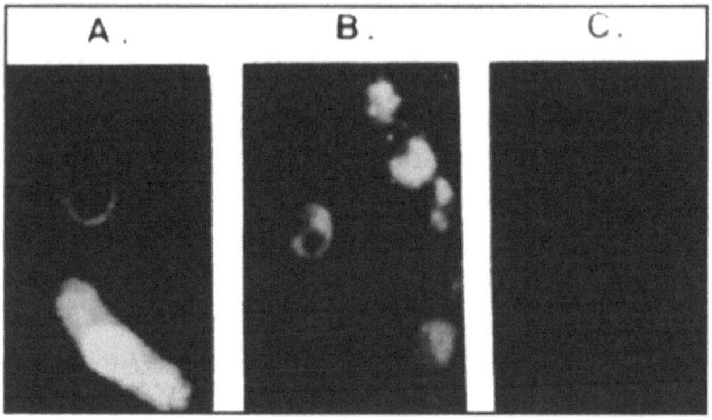

Fig. 8.2. Localization of L1 and L2 capsid proteins in *S. pombe* by indirect immunofluorescence. Fission yeast expressing HPV 6L1/L2 genes werestained with a neutralizing antibody against HPV 11 (B2) (A), with anti-HPV6 L2 (3F7) (B) and with a control antibody (anti-human Rb antibody) (C). Strong L1 protein signals are observed in the nucleus (A). L2 is predominantly stained in the nucleus but spotty staining is also seen in the cytoplasm (B). No signals are observed with a control antibody (C).

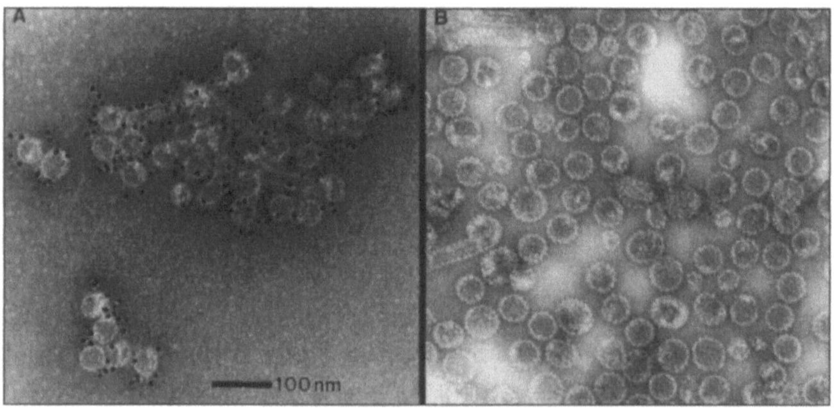

Fig. 8.3. Electron microscopic examination of HPV 16 virus-like particles.
Three microliters of purified VLP samples were anti-HPV16mouse monoclonal antibodies and subsequently with goat anti-mouse immunogold conjugate. The samples were negatively stained with 1% aqueous uranrl acetate.
(A). Immuno-gold labelling with monoclonal antibody, Camvir 6, which recognized HPV16 alone. (B) Immuno-gold labelling with the control monoclonal antibody anti-HPV16-E2 protein.

cytoplasm and membrane.[14] Most of HPV 16-27L1 virus-like particles (VLP) were found in the nuclear fraction of *S. pombe* (Fig. 8.3). Nuclear localization seems to be important for the VLP formation in yeast but does not ensure efficient assembles. The localization pattern of PL1 did not differ from that of 27L1, although the PL1, which was derived from the prototype HPV 16 sequenced by Seedorf et al,[17] did not assemble efficiently. The L2 protein was also localized in the nucleus, but the role of L2 in capsid assembly is unclear, because L1 alone in yeast formed the same VLP as those derived from *S. pombe* containing L1 and L2 genes.[14]

Comparison of VLP Yield in Fission and Budding Yeast

We introduced expression of 27L1 protein in the recombinant *S. cerevisiae* (27L1/C) and tried to purify the VLP by the same method as used for the *S. pombe* containing 27L1(27L1/P). A few HPV16 VLP were purified from lysate of the 27L1/C (Fig. 8.4). In contrast, many virus-like particles of yeast retrotransposon Ty,[18] were accidentally co-purified from *S. cerevisiae*. Ty-VLP appear as 30-45 nm particles containing core DNA structures and were, therefore, distinguishable from the HPV-VLP by careful observation (Fig. 8.4).

We examined yield of the HPV-VLP from different yeast clones by counting mean numbers of VLP from each yeast clone in ten random fields of electron microscopy at x 34,000. Three different *S. pombe* clones and one *S. cerevisiae* clone were compared (Table 8.1). HPV 16-27L1 assembled most efficiently to form virus-like particles in *S. pombe*. Production level of the 27L1-VLP from *S. pombe* (27L1/P) was about 1000 times more than PL1/P (*S. pombe* containing PL1)

Table 8.1. Yield of the virus-like particles in the different recombinant yeasts

Yeast clone	Mean number of VLP [a] (0.5 mg/ml [b] x 34,000)	Efficiency of VLP formation [c]
S. pombe-H6 L1	2250 (0)[d]	750
S. pombe-H16 PL1	3 (0)	1
S. pombe-H16 27L1	2960 (0)	987
S. cerevisiae-H16-27L1	280 (200)	93

a: The mean number of particles in ten fields which was calculated from three different preparations is shown. b: each sample was adjusted by total protein amount of 0.5mg/ml. c: Comparative efficiency of production of the virus-like particles (prototype L1 of HPV 16 is set at 1.0) d: number of the Ty retrotransposon particles.

and 110 times more than 27L1/C. Approximately ten times as many Ty-VLP as 27L1-VLP were purified from *S. cerevisiae.*

Glycosylation and Phosphorylation of the HPV Capsid Protein

Protein modifications are important for protein folding and enzyme activity in many structural or functional proteins. Previous studies showed that HPV16 L1 protein was glycosylated and phosphorylated.[19,20] We studied such modification of the L1 protein in yeast. Lysates of HPV6L1, HPV 16 PL1, 27L1 in *S. pombe* and 27L1 in *S. cerevisiae* were incubated over night with N-glycosidase F and with endoglycosidase F according to the manufacturer's instructions (Fig. 8.5). The samples were run on SDS-PAGE gel and analyzed by immunoblot with Camvir 1. No shift of L1 bands suggesting deglycosylation was observed in any of the samples after the digestion. We also treated these yeast clones with tunicamycin for 20 hours during induction of L1 protein, without any indication of glycosylation of the protein (data not shown). We do not know the reason why our results are different from previous reports,[19,20] although earlier studies used higher eukaryotes like mammalian cells or insect cells for L1 synthesis and this may be relevant.

Next we examined the phosphorylation state of the L1 protein. The *S. cerevisiae* containing 27L1 gene (27L1/C), the *S. pombe* containing 27L1 gene (27L1/P) and *S. pombe* containing PL1 (PL1/P) were labelled with ^{32}P. In an immunoprecipitation assay, a weak L1 band for PL1/P and no bands for 27L1/C and -27L1/P were detected by reaction with Camvir 1, whereas no band for PL1/P and very faint bands for 27L1/C and 27L1/P were observed by reaction with Camvir 6 (Fig.

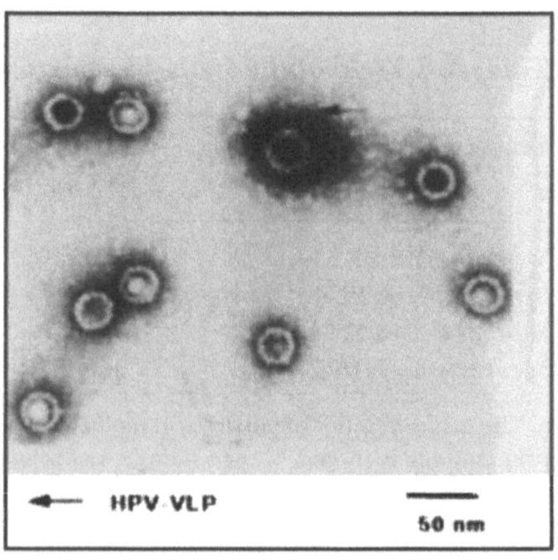

Fig. 8.4. HPV-VLP and Ty-VLP in electron microscopy. The larger particle (about 50 nm) indicated with a black arrow is HPV 16-27L1 VLP and the other smaller particles (20-30 nm) are Ty-VLP. The HPV16 particle alone is reactive to Camvir 6 as HPV-VLP and Ty-VLP in electron microscepy. The larger particle (about 45-50nm), indicated by a black arrow, is HPV 16-27L1 VLP, and the other smaller particles are Ty-VLP. The HPV 16 particle alone is reactive to Camvir 6, as shown by black gold-dots on the VLP surface.

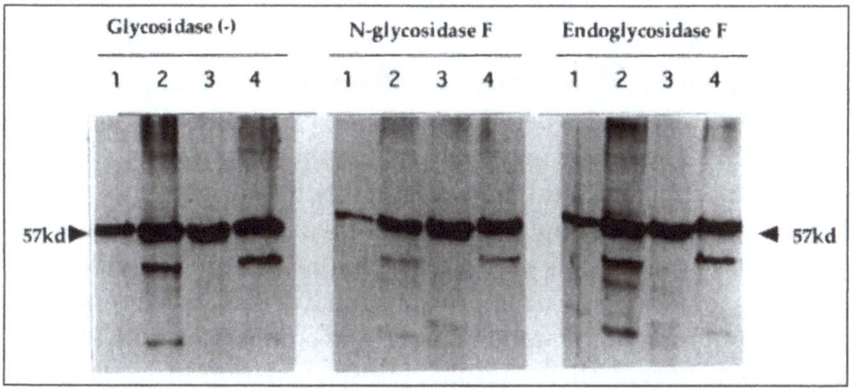

Fig. 8.5. Deglycosylation of HPV 16 L1 protein in different yeast clones.
Lane 1 (HPV 6-L1), lane 2 (HPV 16-PL1) and lane 3 (HPV 16-27L1) represent *S. pombe* clones and lane 4 (HPV 16-27L1) is a *S. cerevisiae* clone. Cell lysate of each yeast clone was incubated over night without enzyme (left) and with N-glycosidase F (SIGMA)(middle), and with endoglycosidase F (SIGMA)(right) in the condition according to the manufacturer's instructions. After the SDS-PAGE, all samples were subjected to immunoblot analysis with anti-HPV 16 capsid antibody, Camvir 1. The largest band is 57 kDa-L1 protein and the smaller bands are degeneration products of L1.

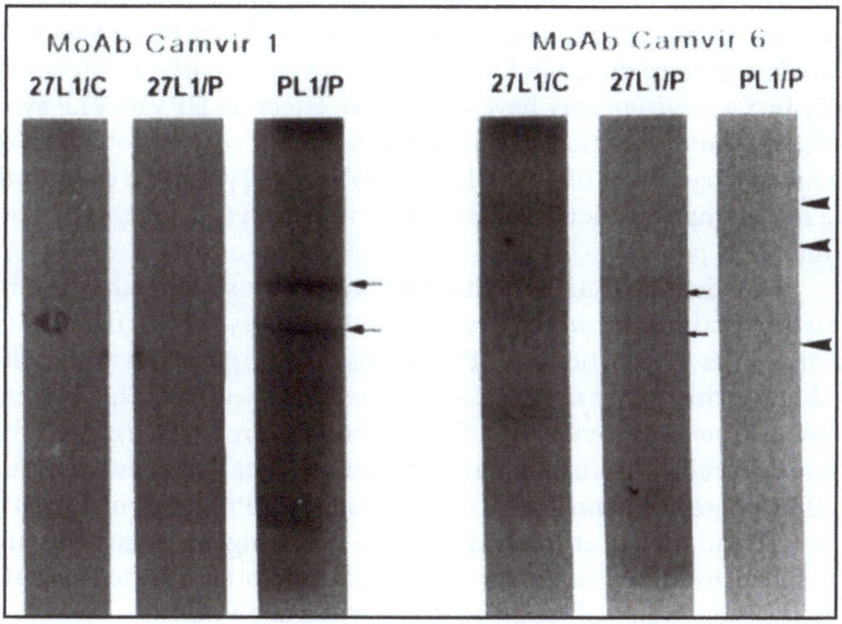

Fig. 8.6. Phosphorylation of the L1 proteins in different yeast clones. The *S. cerevisiae* containing 27L1 gene (27L1/C), the *S. pombe* containing 27L1 gene (27L1/P) and *S. pombe* containing PL1 (PL1/P) were labelled with ^{32}P by incubation in yeast medium containing 300 μCi of inorganic ^{32}P. The labelled proteins were subsequently immunoprecipitated with two anti-L1 antibodies, Camvir 1 and Camvir 6, run on SDS-PAGE gel, and the dried gel was exposed on X-ray film.

8.6). The results show that a phosphorylated -PL1 band was detected only when the PL1 was incubated with Camvir 1 but not with Camvir 6. Camvir 1 is known to react to denatured HPV 16 L1 protein but not to intact HPV 16-VLP, whereas Camvir 6 is able to react to the intact VLP.[11] The results may suggest that phosphorylated L1 protein does not participate in the capsid assembly, because the phosphorylated PL1 was not precipitated with Camvir 6 and, in addition, the phosphorylation signals were weakest in the 27L1 protein which was most efficiently assembled. A previous report also showed very weak phosphorylation in the same L1 protein in comparison with HPV 16 L2 protein.[19] From all the results, phosphorylation of L1 does not seem to be important for self-assembly of L1 protein into VLP.

Conclusion

We obtained no clear evidence of glycosylation and phosphorylation of HPV 16 L1 proteins synthesized in yeast, suggesting that such protein modification is not important for effective assembly of the protein. As we reported before,[13] an amino acid mutation in position 202 in PL1 (Fig. 8.1) is probably crucial to the poor self-assembly of this mutant protein. However, it is not clear why HPV16-VLP are less well produced in *S. cerevisiae* than in *S. pombe*. Existence of Ty in *S. cerevisiae* may have an adverse effect on HPV16-VLP synthesis. If some functions facilitating assembly of capsid proteins were shared in synthesis of both VLP of HPV 16 and Ty, it could constitute a reason that production of HPV-VLP in *S. cerevisiae* is less efficient than *in S. pombe.*

We demonstrate here that HPV capsid protein synthesized in fission yeast *Schizosaccharomyces pombe* can assemble efficiently to form virus-like particles (VLP) and this microorganism is more suitable for this purpose than *Saccharomyces cerevisiae*. This system using *S. pombe* provides an ideal method for production of VLP for use as vaccines against human papillomaviruses and as antigens for ELISA detecting anti-HPV capsid antibody.[11] Problems of toxicity, purity and high cost in other systems using higher eukaryotes are avoided and the system may be applicable to other viruses for particle production and for structural protein analysis.

Acknowledgments

I would like to specially thank Massimo Tommasino (Cancer Institute, German) for providing me with the yeast expression system and giving me precious suggestions, and, also thank Lionel

Crawford (Imperial Cancer Research Fund, UK) and Masaki Inoue (Kanazawa University, Japan) for supervision in all my works, Dr. Margaret Stanley (Department of Pathology, University of Cambridge, UK) for providing the monoclonal antibody Camvir 6 and Dr. N. Christensen (Hershey Medical Center, MA, USA) for providing the neutralizing monoclonal antibody B2 against HPV 11.

References

1. Durst M, Gissman L, Ikenberg H et al. A papillomavirus from cervical carcinoma and its prevalence in cervical cancer biopsy samples from different geographical regions. Proc Nat Acad Sci USA 1983; 80:3812-3815.
2. Boshart M, Gissmann L, Ikenberg H et al. A new type of papillomavirus DNA, its presence in genital cancer biopsies and in cell lines derived from cervical cancer. EMBO J 1984; 3:1151-1157.
3. zur Hauzen H. Human papillomaviruses in the pathogenesis of anogenital cancer. Virology 1991; 184:9-13.
4. Strong M, Vaughan C, Healy G. In: Healy GB, McGill T, J-I, eds. Laryngo-tracheal problems in the pediatric patients. Springfield, IL 1979:88-98.
5. Finch JT, Klug A, The structure of viruses of the papilloma-polyoma type. III. Structure of rabbit papilloma virus. J Mol Biol 1965; 13:1-12.
6. Zhou J, Sun XY, Stenzel DJ et al. Expression of vaccinia recombinant HPV 16 L1 and L2 proteins in epithelial cells is sufficient for assembly of HPV virion-like particles. Virology 1991; 185:251-257.
7. Hagensee ME, Yaegashi N, Galloway DA. Self-assembly of human papillomavirus type 1 capsids by expression of the L1 protein alone or by coexpression of the L1 and L2 capsid proteins. J Virol 1993; 67:315-322.
8. Rose RC, Bonnez W, Reichman RC et al. Expression of human papillomavirus type 11 L1 protein in insect cells: *in vivo* and *in vitro* assembly of virus-like particles. J Virol 1993; 67:1936-1944.
9. Kirnbauer R, Taub J, Greenstone H, Roden R et al. Efficient self-assembly of human papillomavirus type 16 L1 and L1-L2 into virus-like particles. J Virol 1993; 67:6929-6936.
10. Volpers C, Schirmacher P, Streeck RE et al. Assembly of the major and minor capsid protein of human papillomavirus type 33 into virus-like particles and tubular structures in insect cells. Virology 1994; 200:504-512.
11. Sasagawa T, Inoue M, Lehtinen M et al. Serological responses to human papillomavirus (HPV) types 6 and 16 virus-like particles (VLP) in patients with cervical neoplastic lesions. Clinic Diag Labo Immunology 1996; 4:403-410.
12. Pushko P, Sasagawa T, Cuzick J et al. Sequence variation in the capsid protein genes of HPV 16. J Gen Virol 1994; 75:911-916.

13. Maundrell K. Thiamine-repressible expression vectors pREP and pRIP for fission yeast. Gene 1993; 123:127-130.
14. Sasagawa T, Pushko P, Steer G, Gschmeissner SE et al. Synthesis and assembly of virus-like particles of human papillomavirus type 6 and type 16 in fission yeast Schizosaccharomyces pombe. Virology 1995; 206:126-135.
15. Baldari C, Murray JAH, Cesareni G et al. A novel leader peptide which allows efficient secretion of a fragment of human interleukin 1B in *Saccharomyces cerevisiae.* EMBO J 1987; 6:229-234.
16. Zhou J, Doorbar J, Sun XY et al. Identification of the nuclear localization signal of human papillomavirus type 16 L1 protein. Virology 1991;185:625-632.
17. Seedorf K, Krammer G, Durst M et al. Human papillomavirus type 16 DNA sequence. Virology 1985; 145:181-185.
18. Kingsman AJ, Kingsman SM. Ty: A retroelement moving forward. Cell; 53:333-335.
19. Xi SZ, Banks LM. Baculovirus expression of the human papillomavirus type 16 capsid proteins: detection of L1-L2 protein complexes. J Gen Virol 1991; 72:2981-2988.
20. Zhou J, Sun XY, Stenzel DJ et al. Expression of vaccinia recombinant HPV 16 L1 and L2 ORF proteins in epithelial cells is sufficient for assembly of HPV virus-like particles. Virology 1991; 185:251-257.

CHAPTER 9

Expression, Purification and Analyses of Cell-Cycle Regulatory Proteins in *S. pombe*

Kazuhiro Shiozaki and Paul Russell

Introduction

The fission yeast *Schizosaccharomyces pombe* has long been recognized as one of the most useful model systems for investigating cell cycle regulation.[1] *S. pombe* has been favored as an experimental organism largely because of the power of yeast genetics and molecular biology and because fission yeast has been repeatedly proven to serve as a useful paradigm for uncovering the key elements of cell control. Indeed, a large number of the critical components of cell cycle control, such as cdc2 kinase, Wee1 kinase and cdc25 phosphatase, were first discovered in fission yeast. Although major technical advances have recently made it possible to carry out significant functional in vivo analyses of cell cycle control proteins in more complex organisms such as the fruit fly and mice, fission yeast continues to serve as one of leading experimental systems for elucidating central features of eukaryotic cell cycle control.[2]

The advantages in using *S. pombe* as an experimental tool for cell cycle studies can be summarized as follows:

1) Since *S. pombe* cells normally exist in the haploid state, recessive mutations can be isolated. A number of genes with a function at a particular point of the cell cycle were identified as *cdc* mutations, which arrest cell cycle progression without affecting cell growth. This results in a highly elongated cell

Foreign Gene Expression in Fission Yeast Schizosaccharomyces pombe, edited by Yuko Giga-Hama and Hiromichi Kumagai.

morphology in *S. pombe*, that can be easily identified by visual screening.

2) Various molecular genetic techniques have been developed; genes can be cloned by complementation of mutant phenotypes. Using homologous recombination, the genomic copy of a gene can be replaced with a gene fragment that has been modified in vitro. In addition, vectors with a range of expression levels are available, which allows another set of genetic and biochemical experiments including expression of foreign genes in *S. pombe*.[3]
3) The ability to synchronize cells in a culture at a particular point in the cell cycle. There are two methods to synchronize *S. pombe* cells. One approach is to select early G2 cells from an asynchronous culture using elutriation centrifugation. This method separates cells on the basis of size and is favored because it involves virtually no disturbance of cell physiology. The other popular method is to use *cdc* mutations which arrest the cell cycle at a particular stage. *cdc10* mutations are often used to arrest in G1, *cdc22* for S phase arrest, *cdc2* and *cdc25* mutations for G2 arrest, and *nda3* and *nuc2* for M phase arrest.[4-8] These are all conditional mutations and cell cycle-specific arrest can be easily performed by shifting the culture from a permissive temperature to restrictive temperature. Because of its reversible nature, *cdc25-22* mutation is also used for arrest/release experiments; shifting to the restrictive temperature 36°C arrests *cdc25-22* cells at G2/M boundary. By shifting down to the permissive temperature, 25°C, cells resume the cell cycle in a highly synchronous manner.[9,10]

Many of these advantages also make *S. pombe* amenable as a protein expression system. Genes of *S. pombe* and other organisms have been expressed at various levels in fission yeast. By using the synchronization techniques mentioned above, expressed proteins can be recovered with cell-cycle specific modifications. In addition, some mutations of host yeast strains make it possible to high levels of express proteins that would normally be toxic to the cell. In this paper we will introduce two protein expression/purification systems in *S. pombe* that have been frequently and successfully used in our laboratory. While the examples described here involve only *S. pombe* genes, these methods can be easily applied to produce and purify foreign proteins using fission yeast as a host.

Expression of *S. pombe* Genes in the Genome as a Fusion with an Epitope and Six Histidine Residues

This method is used to analyze protein products of *S. pombe* genes without requiring the production of protein specific antibodies. Proteins are expressed as a fusion with a monoclonal antibody epitope and six consecutive histidine residues, which allow detection and purification of the proteins. By integrating the fusion gene construct into the chromosomal locus, expression is regulated from the endogenous promoter in the correct chromatin, so that cell physiology is not disturbed by protein overexpression.

Strain Construction

Strain construction for this method is schematically shown in Figure 9.1.a. A DNA fragment encoding the carboxyl terminus of the protein is cloned into a vector containing the *ura4*$^+$ gene as a selectable marker but no autonomous replication sequences (*ars*). The cloned gene fragment should be longer than ~500 bp in order to

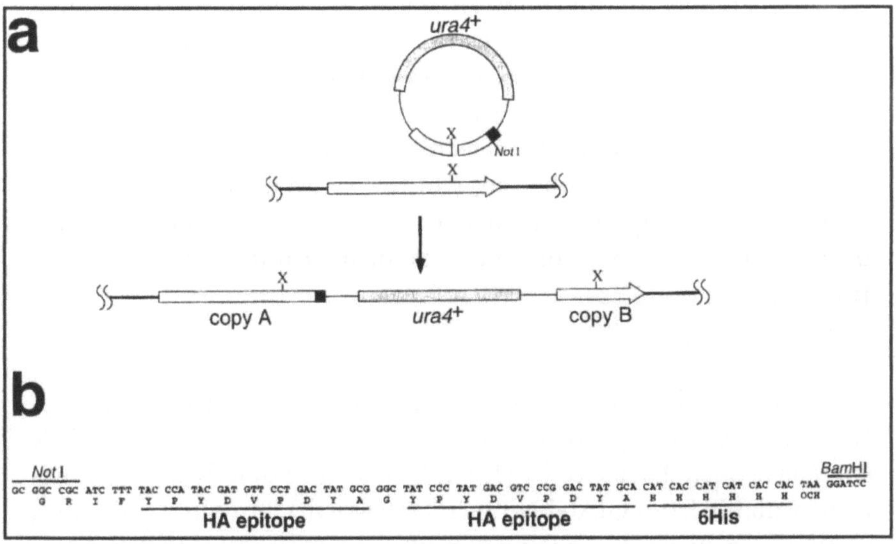

Fig. 9.1. (a) Construction of a strain expressing epitope/6xHis tagged protein. The 3' portion of the ORF of the gene is cloned in a plasmid with *ura4*$^+$ as a selectable marker. In this plasmid, the 3' end of the ORF is further extended in frame by two copies of HA epitope and six consecutive histidine residues, which are followed by a transcriptional terminator sequence of the *nmt1*$^+$ gene (indicated as a filled box). This construct is linearized by restriction digestion at a unique site X around the middle of the cloned ORF and used for transformation of *S. pombe* strains. By homologous recombination, the linearized plasmid is integrated in the genomic locus of the gene. In the resultant strain, protein is expressed as a fusion with epitope and 6xHis in its C-terminus from copy A which is followed by the *ura4*$^+$ marker gene and a truncated inactive copy B of the gene.
(b) The nucleotide sequence of the "tag" used in (a), encoding two copies of HA epitope and six histidine residues, which is derived from pREP1X6HHa described in Wu and Russell.[21]

obtain a high frequency of homologous recombination. We usually use a PCR technique to get an appropriate fragment with a 3' primer containing a *Not*I restriction site, so that the C-terminus of the open reading frame (ORF) of the gene is fused in frame to the sequence shown in Figure 9.1.b. This sequence contains two copies of influenza hemagglutinin epitope (HA), to which antibodies are commercially available, and six consecutive histidine residues.[11] Proteins tagged with the six consecutive histidine residues ('6xHis') are easily purified from crude cell lysate using Ni-NTA-agarose beads as described below.[12] The tag sequence is followed by a termination codon (Fig. 9.1.b) and a transcriptional terminator sequence from the *S. pombe nmt1*$^{+}$ gene.[13] The resulting plasmid is cleaved at a unique restriction site X in the ORF and used for transformation of a *S. pombe* strain carrying a deletion of the *ura4*$^{+}$ locus.[14] Ura^{+} colonies are further selected for stable transformants by a stability test.[15] Stable transformants are expected to have undergone homologous recombination at restriction site X as shown in Figure 9.1.a, and this should be confirmed by Southern hybridization with genomic DNA purified from the transformants. In this strain, the gene of interest is expressed from its own promoter as a fusion with the C-terminus tag (copy A in Figure 9.1.a), which is followed by the *nmt1*$^{+}$ transcriptional terminator and the vector sequence containing the *ura4*$^{+}$ marker gene. While another copy of the 3' end of the gene remains in the genome (copy B in Figure 9.1.a), it is usually non-functional because of the truncation and lack of a promoter sequence for transcription.

Purification of the 6xHis-Tagged Proteins

The chemical principles and experimental conditions for purification of 6xHis tagged proteins using Ni-NTA agarose (QIAGEN Inc.) have been discussed by Hochuli et al and a detailed brochure is also available from QIAGEN Inc. (Chatsworth).[12] In our experience, purification using denaturing conditions has been quite successful with various proteins, although purification of native proteins has not worked as well. There appear to be two major problems with protein purification using non-denaturing conditions. The first problem is that the yield of 6xHis-tagged protein is often low. This may be explained if the 6xHis tag is not fully exposed, thus hindering the interaction between 6xHis sequence and Ni moiety of the resin. The second problem with purification using non-denaturing conditions is that the preparation is typically not pure. *S. pombe* cell lysates contain abundant endogenous proteins which bind tightly to Ni-NTA-beads under non-denaturing conditions. As a consequence the purification efficiency of the 6xHis-tagged proteins of interest can be quite low, especially when relative abundance of the proteins in the cell is low (which is usually the case when the proteins are expressed from their own promoter). Therefore, we mostly use a purification procedure using highly denaturing conditions. The following is a protocol which has been successfully employed for several different proteins in our laboratory.

1) Harvest ~20 OD (optical density at 600 nm) of yeast cells by filtration or centrifugation. Resuspend in 1 ml of ice cold STOP buffer (50 mM NaF, 10 mM EDTA, 100 mM NaCl), transfer to a 1.5 ml microfuge tube (or transfer filter to a microfuge tube containing 1 ml of STOP buffer), spin and discard supernatant and then immediately freeze the microfuge tube using a liquid nitrogen bath. For time course experiments using synchronized cultures, filtration is preferred because the time required for centrifugation (5 minutes or more) can hinder the detection of changes which occur during narrow windows of the *S. pombe* cell cycle. Lysates of 20 OD cells contain 4-5 mg total protein.
2) Resuspend the cell pellet in 0.2 ml of Buffer G (6 M guanidine hydrochloride, 0.1 M Na-phosphate, 50 mM Tris-HCl, pH 8.0). Add acid washed glass beads (0.45-0.50 mm diameter) to just below the meniscus of the suspension. Buffer G and glass beads are stored at 4°C and cooled to 0°C by incubation on ice prior to use. Vortex tubes in a cold room at 4°C for 5-10 min. Examine the cells under the phase-contrast microscope to make sure that > 90% of the cells are disrupted. Disrupted cells will appear phase dark. Elongated *cdc* mutants are particularly easy to break, whereas small cells derived from *wee* strains or stationary phase cultures are somewhat more resistant to breakage.
3) To recover the cell homogenate, make a pin hole at the bottom of the sample tube by piercing with a hot syringe needle heated in a flame. Pierce the tube while holding it immediately above another microfuge tube, then place the pierced tube in the new tube. Briefly centrifuge the stack of the two tubes in a swing out rotor so that the cell homogenate is collected in the bottom tube, leaving the glass beads in the top tube. Add 0.3 ml of Buffer G to the glass beads and repeat centrifugation, which gives approximately 0.5 ml of the cell lysate in the bottom tube.

4) Centrifuge the cell lysate in a microfuge at ~14,000 rpm for 10 minutes at room temperature to pellet insoluble debris. Recover the supernatant into a fresh microfuge tube and mix with 10 µl bed volume of Ni-NTA agarose beads (QIAGEN) which have been washed beforehand three times with distilled water and then three times with Buffer G. Incubate the mixture of the extract and beads at room temperature for 1 hour on a rocking platform.
5) Pellet the beads by a quick spin in a microfuge, and discard the supernatant. Add 1 ml of Buffer G, mix well, briefly centrifuge and discard supernatant. Repeat this wash three times. We perform a "double-spin" in order to minimize loss of the beads during these washes when using a fixed angle rotor. After a first quick spin, turn the microfuge tube 180° and then perform another quick spin. This procedure reduces the amount of beads that stick to the inside wall of the sample tubes.
6) For analysis by SDS-polyacrylamide gel electrophoresis (SDS-PAGE), wash the Ni-NTA-beads three times further with Buffer T (50 mM Tris-HCl, pH 8.0, 250 mM NaCl) to get rid of guanidine hydrochloride, which is precipitated in the presence of SDS. Add 30 µl of SDS-PAGE loading buffer (2x) containing 10% 2-mercaptoethanol, and incubate for 2 min in a boiling water bath.

Analyses of the Purified Proteins

The purification procedure described above is highly reproducible and effective, permitting proteins of interest to be purified and concentrated in a short and simple protocol. Yields typically approach 100%. Proteins are easily detected either by direct staining methods or by immunoblotting using anti-HA epitope antibodies. This strategy has been particularly useful for studying proteins that are present at very low levels in the cell, obviating the need to overproduce proteins or use other extraordinary measures to study low abundance proteins. The method is also ideally suited for studying proteins that are sensitive to proteolysis or dephosphorylation during the purification, since suspension in the extraction buffer containing guanidine hydrochloride causes immediate inactivation of all proteases and phosphatases. In fact, it is probably unnecessary to perform the extraction at low temperature. The method combines the advantages of immunoprecipitation while avoiding some of the problems associated with use of antibody mediated purification methods. In particular, it is compatible with all post-purification immunoblotting procedures, including the use of anti-phosphotyrosine antibodies that are used to monitor tyrosine phosphorylation of proteins. A successful example of this approach is the analysis of Spc1 (Sty1), a stress-activated MAP kinase homolog.[16] A strain expressing Spc1HA6His constructed as above was used to examine the tyrosine phosphorylation of Spc1 along a time course after *S. pombe* cells were subjected to various forms of stress.[16-18] Changes in the amount and phosphorylation state of the $p34^{cdc2}$-inhibitor kinase Wee1, and Nim1 protein kinase which inhibits Wee1, were also examined by similar experimental procedures (R Aligue, L Wu, PR, in preparation).[19,20] While Wee1 and Nim1 are not tyrosine-phosphorylated, mobility shifts in SDS-PAGE can be used to evaluate their phosphorylation on serine/threonine residues.[21] In such experiments we compare the amount and phosphorylation state of the proteins from the samples at various time points; therefore the same amount of cell lysates should be used for each sample. We measure the protein concentration of samples by Bradford's method after the centrifugation in the step 4 above, and then use same amount of proteins for each precipitation.

Modifications

Various modifications of this method are possible depending on the purpose of experiments. Epitope tags other than HA can be used, for example the *myc* epitope has been used for a number of different applications.[22,23] The copy number of the epitope affects the sensitivity for detection, in some constructs we have used three tandem copies of the HA epitope in order to detect proteins that are expressed at very low levels. The position of the epitope/6xHis tag is important in some proteins. We have observed that at least one protein, cdc25, is inactivated by a C-terminal tag. In such cases, tagging at the N-terminus should be tried.

Secondly, the *ura4*$^+$ selectable marker of the integration plasmid may be replaced with other marker genes: the *Saccharomyces cerevisiae LEU2* gene is commonly used to complement the *S. pombe leu1-32* mutation, and the *S. pombe his7*$^+$ gene is used to complement *his7-366* strains.[24] However, *leu1-32* strains carrying a single integrated copy of *LEU2* have a significantly reduced growth rate in selective media which makes it rather difficult to select stable integrants after transformation.[19] On the other hand, stable integrants of *ura4*$^+$ marker gene often grow faster than unstable transformants. For more discussion about selectable markers in *S. pombe*, see Russell.[25]

High-Level Expression of GST-Fusion Proteins in *S. pombe* Using the *nmt1*$^+$ Promoter

S. pombe nmt1$^+$ (no message in thiamine) was identified as a gene whose message is highly induced in growth media lacking thiamine (vitamin B1) and repressed in the presence of thiamine.[26] The *nmt1*$^+$ gene product is thought to be involved in thiamine biosynthesis. *nmt1*$^+$ is highly transcribed under induced conditions, producing 50~100 times more mRNA than the *cyc1*$^+$ gene encoding cytochrome.[27] A series of pREP expression vectors have been constructed using the promoter of this gene.[13,28] Experiments using *lacZ* as a reporter gene indicate that induced expression level from the *nmt1*$^+$ promoter is about twice as much as that from the *adh1* promoter that has been widely used for overexpression of genes in *S. pombe*.[3] Although low basal levels of expression have been detected even in the presence of thiamine, a very high induction ratio (induced/repressed = ~300) allows overexpression of genes whose product is toxic to the cells when overproduced. Induced expression from the *nmt1*$^+$ promoter starts about 12 hours after thiamine depletion and reaches a maximum level within at approximately 16 hours.

The system introduced here utilizes this inducible *nmt1*$^+$ promoter to express proteins as a fusion with glutathione *S*-transferase (GST) at their N-terminus. Expression of GST-fusion proteins in bacteria has been reported and widely used because GST-fusion proteins are easily purified in a single step using affinity chromatography on glutathione agarose under non-denaturing conditions.[29,30] However, high-level expression in *E. coli* often creates problems with protein solubility and most of the expressed proteins form inclusion bodies which need to be solubilized by denaturing reagents.[31] In contrast, almost all of the proteins overexpressed in *S. pombe* remain soluble, therefore they can be easily purified in a biologically active form. Moreover, proteins expressed in yeast may undergo modifications such as phosphorylation that can be essential for activity. These features of this system are ideal to prepare proteins as materials for various biochemical experiments.

pREP1-KZ Expression Vector

An expression plasmid to produce GST-fusion proteins in *S. pombe* was constructed by modifying the pREP1 vector containing the *nmt1*⁺ promoter (pREP1-KZ, Fig. 9.2.a).[26] The GST gene and following linker sequence from the pGEX-KG vector was ligated downstream of the *nmt1*⁺ promoter.[30] This linker sequence encodes the thrombin cleavage site (Leu-Val-Pro-Arg-Gly-Ser) and a "kinker" composed of five consecutive glycine residues which facilitates thrombin cleavage (Fig. 9.2.b). The insert cDNA is followed by a short sequence between *Not*I and *Bam*HI sites encoding two copies of HA epitope and 6xHis (Fig. 9.1.b) which tag the GST fusion protein at the C-terminus. We usually amplify gene fragments by PCR using a 5' primer with a *Nde*I site and a 3' primer with a *Not*I site, and clone them between the *Nde*I and *Not*I sites of the vector. When the C-terminal tag is not required, termination codons are introduced in the 3' primer, or insert cDNAs are cloned in frame using the *Bam*HI or *Sma*I sites of the vector. The latter method eliminates the glycine

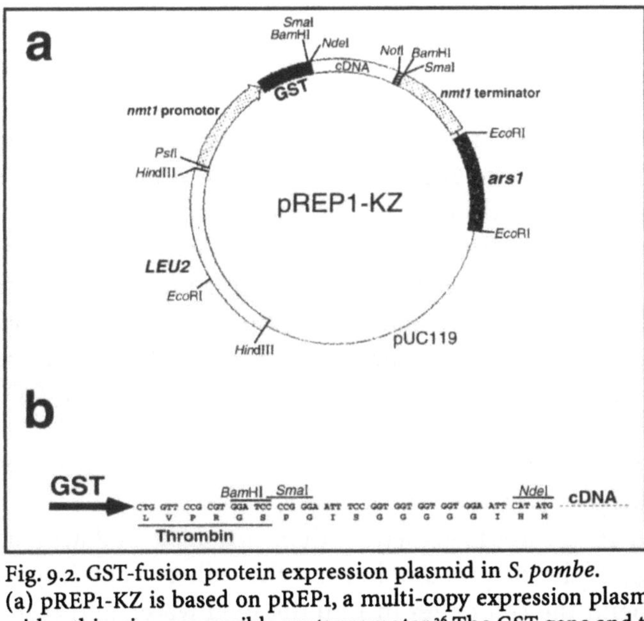

Fig. 9.2. GST-fusion protein expression plasmid in *S. pombe*.
(a) pREP1-KZ is based on pREP1, a multi-copy expression plasmid with a thiamine-repressible *nmt1* promoter.[26] The GST gene and the following linker were derived from pGEX-KG vector for bacterial expression of GST-fusion protein.[30] cDNAs are cloned between *Nde*I and *Not*I sites which are followed by a sequence encoding two copies of HA epitope and 6xHis (hatched box, see Fig. 9.1(b) for sequence).
(b) The nucleotide sequence of the linker following the GST encoding region (arrow). The linker contains thrombin cleavage site and a "kinker" composed of five consecutive glycine residues. *Bam*HI, *Sma*I and *Nde*I sites are included for cloning, although *Bam*HI and *Sma*I are not unique in the plasmid, and digestion by either of these enzymes will cleave off the sequence for the C-terminal tag (see the map in a).

kinker. pREP1-KZ replicates and exists as a episomal multi-copy plasmid by virtue of the *ars1* sequence, and contains *LEU2* as a selectable marker.

Induction and Purification of GST-Fusion Proteins

In our laboratory several proteins of different cellular functions have been expressed using the pREP1-KZ vector described above: the MAP kinase homolog Spc1 and its activator Wis1 MAP kinase kinase, Pyp1/Pyp2 tyrosine phosphatases, the Orp2 subunit of the *S. pombe* origin recognition complex and cdc25 phosphatase (R Soni and P R, unpublished results).[16,17,32] The following is a protocol for the GST-Spc1 protein, but it is applicable to other GST-fusion proteins with minor modifications (see below).[16]

1) Transform a *S. pombe* wild-type strain (*h⁻ leu1-32*) with the plasmid pREP1-KZ-Spc1 and grow the transformants on a minimal medium (EMM2) agar plate containing 1 µM thiamine.
2) Set up a 5ml preculture of the transformant in liquid EMM2 supplemented with 1 µM thiamine and grow for 8~12 hours at 32°C with shaking.
3) Harvest the cells from the preculture by centrifugation in a sterile tube. Suspend the cells into fresh EMM2 medium without thiamine, spin down the cells again, and discard the supernatant. Repeat this washing three times to eliminate trace amounts of thiamine. Resuspend the cells into 10 ml of EMM2 medium and measure the OD at 600 nm. Inoculate an appropriate volume of this cell suspension into 50 ml of EMM2 so that the cell density after 16 to 20-hours growth will be 0.5-1.0 OD_{600} (The doubling time of wild-type *S. pombe* cells in EMM2 media is 2.5-3 hours at 32°C). Grow the culture overnight at 32°C with shaking.
4) After 16~20 hours growth in thiamine-free media, measure the OD_{600} of the culture and harvest the cells by centrifugation. Resuspend the cell pellet in 1 ml distilled water, and transfer to a microfuge tube a volume of the cell suspension equivalent to 20 OD_{600}. Spin down the cells and discard the supernatant. The cell pellet can be kept in a -80°C freezer after being frozen in liquid nitrogen, or processed immediately as described below.
5) The following procedures are performed at 4°C. Suspend the cell pellet in 200 µl of L buffer (50 mM Tris-HCl, pH 8.0, 400 mM NaCl, 1 mM EDTA, 1 mM 2-mercaptoethanol, 10% glycerol, 50 mM NaF, 1 mM Na3VO4, 1 mM PMSF) and add glass beads (0.45-0.50 mm) to just below the meniscus of the sus-

pension. Vortex vigorously for 5~10 min and examine the cells under the phase-contrast microscope to make sure that >90% of the cells are disrupted.

6) Separate the cell homogenate from glass beads as described in step 3, p. 137. Dilute the recovered homogenate with equal volume (~200 µl) of L buffer containing 0.2% NP-40. Centrifuge the sample at top speed for 10 minutes in a microfuge to eliminate insoluble debris. Recover the supernatant carefully and add 20 µl (bed volume) of Glutathione Sepharose 4B (Pharmacia Biotech) pre-equilibrated in L buffer containing 0.1% NP-40. Incubate for 1 hour on a rocking platform.
7) Spin down the Sepharose beads and discard the cell lysate. Wash the beads four times with L buffer containing 0.1% NP-40. Take 1/4 (~5 µl) of the beads and boil in 20 µl of SDS-PAGE loading buffer (2x) containing 10% 2-mercaptoethanol to check the purification of the GST-Spc1 by SDS-PAGE.

By this procedure, 2-4 µg of GST-Spc1 protein is purified from 20 OD_{600} cells (~30 ml culture equivalent). Yield varies dramatically depending on the protein. Examples of GST-fusion proteins purified from *S. pombe* cells are shown in Figure 9.3.

For some proteins, minor modifications of the protocol might be required for maximum activity and yield. First, the kinetics of the accumulation of induced GST-fusion proteins after thiamine-depletion may be different between proteins. It is recommended that a pilot experiment be performed to monitor the expression level of the GST fusion protein during a 12 to 24 hour time course by probing crude cell lysates by anti-HA or anti-GST antibodies. Second, in order to retain activity of the GST-fusion proteins during the

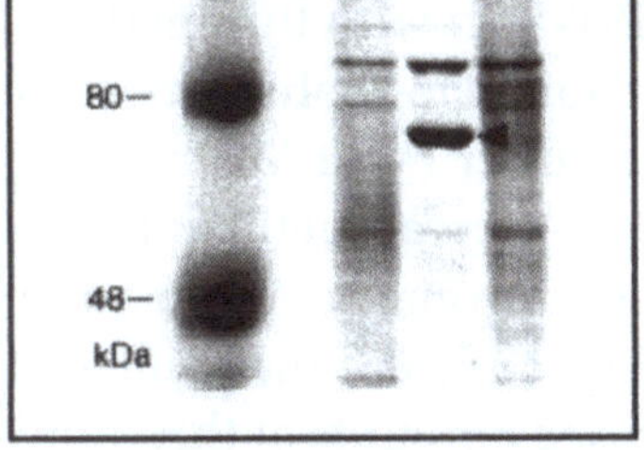

Fig. 9.3. Components of the stress-activated signal transduction pathway in *S. pombe*[16,34,18] expressed as GST-fusions. Wis1 MAP kinase kinase (lane 1), Spc1 MAP kinase (lane 2) and Atf1 transcription factor (lane 3) were expressed using pREP1-KZ plasmid. After purification by glutathione-Sepharose, protein equivalents of 5 OD600 cells were analyzed by 7.5% SDS-PAGE followed by Coomassie blue staining. Purified proteins are indicated by arrowheads. GST-Wis1 protein ran diffusing as two bands, presumably because of phosphorylation or partial proteolysis. Lane M is molecular weight markers.

purification procedure, the composition of the buffers should be also considered. For example, L buffer in the protocol above contains inhibitors of protein phosphatases (NaF and Na_3VO_4) to prevent dephosphorylation of GST-Spc1, which abolishes the protein kinase activity of Spc1.[16] However, these reagents would not be appropriate in buffers used for purification of GST-fusion phosphatases.

Analyses of Purified GST-Fusion Proteins

GST-fusion proteins purified on glutathione-Sepharose beads can be used immediately for various experiments, since most proteins can be employed for biochemical reactions with their GST moiety bound to the beads. Because of stable binding between GST and glutathione, buffers are easily exchanged without losing bound proteins by washing the beads 1 to 3 times with the buffer appropriate for further experiments. After the purification procedure described above, protein kinase activity of GST-Spc1 was assayed by incubating the Sepharose beads with a substrate protein and [γ-^{32}P]ATP following washes the glutathione beads with kinase reaction buffer.[16]

Glutathione-Sepharose purification of GST-fusion proteins is also useful to detect protein-protein interactions in *S. pombe* cells. Proteins coprecipitated with GST-fusion proteins on Sepharose beads are analyzed by SDS-PAGE, followed by immunoblotting or other methods for detection. In vivo interactions between GST-Spc1 and Pyp1 tyrosine phosphatase, between GST-Pyp2 tyrosine phosphatase and Spc1 and, between GST-Atf1 transcription factor and Spc1 have been detected by this procedure.[16-18] Unfused GST purified from cells transformed with pREP1-KZ serves as an appropriate control for these experiments.

It is also possible to elute fusion proteins from the Sepharose beads using buffer containing glutathione as reported.[30] After elution, the GST moiety of the fusion proteins can be cleaved off by thrombin, if required. Alternatively, thrombin digestion can be performed while the protein is still bound to glutathione-Sepharose.[30] Figure 9.4 shows thrombin digestion of GST-Spc1 with this method. After step 7 on the previous page, GST-Spc1 bound to Sepharose beads was washed once with thrombin cleavage buffer (50 mM Tris-HCl, pH 8.0, 150 mM NaCl, 2.5 mM $CaCl_2$, 0.1% 2-mercaptoethanol) and resuspended in the same buffer containing thrombin. About 50 to 60% of the Spc1 was cleaved from GST (Fig. 9.4), and the yield was not affected by longer incubation or increased amounts of thrombin

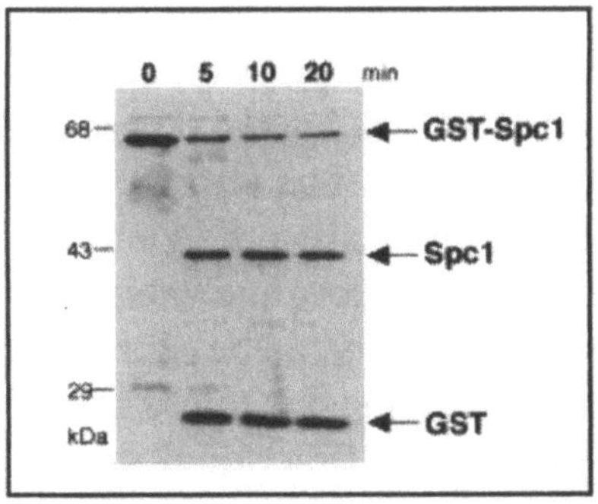

Fig. 9.4. Thrombin digestion of GST-Spc1 fusion protein. GST-Spc1 fusion protein purified on glutathione-Sepharose beads was washed once thrombin digestion buffer and divided into four aliquots. Each aliquot was resuspended in 30 ml of the same buffer containing 0.1 unit of thrombin (Sigma), and incubated at room temperature. Digestion of each aliquot was terminated at 0, 5, 10, and 20 min, respectively, by adding 2.5 mM EGTA and 1 mM PMSF, and analyzed by 12.5% SDS-PAGE followed by coomassie blue staining.

in the reaction (data not shown). Digestion efficiency appears to be different from protein to protein; in an experiment under the same conditions ~100% of GST-Wis1 was cleaved within 10 minutes (data not shown). Pilot experiments are recommended with a range of incubation time and amount of thrombin. However, these additional steps of elution or thrombin cleavage may lose some portion of proteins because of non-specific binding of the proteins to Sepharose beads or incomplete digestion.

Modifications

One of the advantages of this system is that GST-fusion proteins can be expressed in strains of various genetic backgrounds, by simply using different *S. pombe* strains for transformation. This allows one to purify the same protein with different states of modification, or some of the proteins can be expressed at a high level only in particular mutant strains. GST-Spc1 is an example of the former case. GST-Spc1 protein purified wild-type cells stressed by high-osmolarity is phosphorylated on threonine-171 and tyrosine-173 and active as a protein kinase.[16] On the other hand, in a mutant strain lacking functional *wis1*$^{+}$ gene, which encodes a MAP kinase kinase homolog responsible for Spc1 phosphorylation, GST-Spc1 is not subjected to activating phosphorylations and serves as a good substrate for in vitro assays of Wis1 activity. GST-Wis1 and GST-cdc25 proteins fall in the latter case. Wis1 and cdc25 are activators of Spc1 and cdc2, respectively, and overexpression of Wis1 or cdc25 from the *nmt1*$^{+}$ promoter result in cell lethality or a poor growth rate.[16,33] Therefore, a *spc1*$^{-}$ mutant strain was used to produce GST-Wis1,[16] and *cdc2*$^{-}$ mutants for GST-cdc25 production (R Soni and P R, unpublished result).

Conclusions

We have had outstanding success with the expression and purification of functional GST fusion proteins in *S. pombe.* There are many advantages to the system. All the proteins thus far expressed have been soluble and functional, and in most cases the yield has been high. In general yeasts are quite easy to grow and maintain, and the growth media is inexpensive. The proteins can be expressed in a wide variety of mutant backgrounds, allowing one to use genetics to confirm that the GST fusion proteins are functional in vivo and to control protein modification. This system will be also useful for production of proteins from other organisms.

Acknowledgments

We thank Janet Leatherwood, Lin Wu, and members of P. Russell's laboratory at The Scripps Research Institute for advice and discussions, and Mitsue Shiozaki for technical assistance. K. S. was supported by fellowships from the Human Frontier Science Program and California Division-American Cancer Society (#1-6-95). This work was supported by an NIH grant awarded to P. R.

References

1. Hayles J, Nurse P. Genetics of the fission yeast *Schizosaccharomyces pombe.* Annu Rev Genet 1992; 26:373-402.
2. Nurse P. Universal control mechanism regulating onset of M-phase. Nature 1990; 344:503-8.
3. Forsburg SL. Comparison of *Schizosaccharomyces pombe* expression systems. Nuc Acids Res 1993; 21:2955-6.
4. Nurse P, Bisset Y. Gene required in G1 for commitment to the cell cycle and G2 for control of mitosis in fission yeast. Nature 1981; 292:558-60.
5. Russell P, Nurse P. *cdc25*$^{+}$ functions as an inducer in the mitotic control of fission yeast. Cell 1986; 45:145-53.
6. Gordon CB, Campbell JL. A cell cycle-responsive transcriptional control element and a negative control element in the gene encoding DNA polymerase a in *Saccharomyces cerevisiae.* Proc Natl Acad Sci USA 1991; 88:6058-62.
7. Hiraoka Y, Toda T, Yanagida M. The NDA3 gene of fission yeast encodes β-tubulin: a cold-sensitive *nda3* mutation reversibly blocks spindle formation and chromosome movement in mitosis. 1984; 39:349-58.
8. Hirano T, Hiraoka Y, Yanagida M. A temperature-sensitive mutation of the *Schizosaccharomyces pombe* gene *nuc2*$^{+}$ that encodes a nuclear scaffold-like protein blocks spindle elongation in mitotic anaphase. J Cell Biol 1988; 106:1171-83.

9. Booher RN, Alfa CE, Hyams JS et al. The fission yeast cdc2/cdc13/suc1 protein kinase: regulation of catalytic activity and nuclear localization. Cell 1989; 58:485-97.
10. Moreno S, Hayles J, Nurse P. Regulation of $p34^{cdc2}$ protein kinase during mitosis. Cell 1989; 58:361-72.
11. Field J, Nikawa J, Broek D et al. Purification of a RAS-responsive adenyl cyclase complex from *Saccharomyces cerevisiae* by use of an epitope addition method. Mol Cell Biol 1988; 8:2159-65.
12. Hochuli E, Dobeli H, Schacher A. New metal chelate adsorbents selective for proteins and peptide containing neighboring histidine residues. J Chrom 1987; 411:177-84.
13. Maundrell K. Thiamine-repressible expression vectors pREP and pRIP for fission yeast. Gene 1993; 123:127-30.
14. Grimm C, Kohli J, Murray J et al. Genetic engineering of *Schizosaccharomyces pombe*: A system for gene disruption and replacement using the *ura4* gene as a selectable marker. Mol Gen Genet 1988; 215:81-6.
15. Moreno S, Klar A, Nurse P. Molecular genetic analysis of fission yeast *Schizosaccharomyces pombe*. Meth Enzymol 1991; 194:795-823.
16. Shiozaki K, Russell P. Cell-cycle control linked to the extracellular environment by MAP kinase pathway in fission yeast. Nature 1995; 378:739-43.
17. Degols G, Shiozaki K, Russell P. Activation and regulation of the Spc1 stress-activated protein kinase in *Schizosaccharomyces pombe*. Mol Cell Biol 1996; 26:2870-77.
18. Shiozaki K, Russell P. Conjugation, meiosis, and the osmotic stress response are regulated by Spc1 kinase through Atf1 transcription factor in fission yeast. Genes Dev 1996; 10:2276-88.
19. Russell P, Nurse P. Negative regulation of mitosis by *wee1*$^{+}$, a gene encoding a protein kinase homolog. Cell 1987; 49:559-67.
20. Russell P, Nurse P. The mitotic inducer *nim1*$^{+}$ functions in a regulatory network of protein kinase homologs controlling the initiation of mitosis. Cell 1987; 49:569-76.
21. Wu L, Russell P. Nim1 kinase promotes mitosis by inactivating Wee1 tyrosine kinase. Nature 1993; 363:738-41.
22. Evan GI, Lewis GK, Ramsay G et al. Isolation of monoclonal antibodies specific for human c-*myc* proto-oncogene product. Mol Cell Biol 1985; 5:3610-6.
23. Munro S, Pelham HRB. Use of peptide tagging to detect proteins expressed from cloned genes: deletion mapping functional domains of *Drosophila* hsp70. EMBO J 1984; 3:3087-93.
24. Apolinario E, Nocero M, Jin M et al. Cloning and manipulation of the *Schizosaccaharomyces pombe his7*$^{+}$ gene as a new selectable marker for molecular genetic studies. Curr Genet 1993; 24:491-5.
25. Russell P. Gene Cloning and expression in fission yeast. In: Nasim A, Young P, Johnson BF, ed. Molecular Biology of the Fission Yeast.

Cell Biology: A Series of Monographs. San Diego: Academic Press, 1989:243-71.
26. Maundrell K. *nmt1* of fission yeast. J Biol Chem 1990; 265:10857-64.
27. Russell PR, Hall BD. Structure of the *Schizosaccharomyces pombe* cytochrome *c* gene. Mol Cell Biol 1982; 2:106-16.
28. Basi G, Schmid E, Maundrell K. TATA box mutations in the *Schizosaccharomyces pombe nmt1* promoter affect transcription efficiency but not the transcription start point or thiamine repressibility. Gene 1993; 123:131-6.
29. Smith DB, Johnson KS. Single-step purification of polypeptides expressed in *Escherichia coli* as fusions with glutathione *S*-transferase. Gene 1988; 67:31-40.
30. Guan K, Dixon JE. Eukaryotic proteins expressed in *Escherichia coli*: An improved thrombin cleavage and purification procedure of fusion proteins with Glutathione *S*-Transferase. Anal Biochem 1991; 192:262-7.
31. Watt RA, Shatzman AR, Rosenberg M. Expression and characterization of the human c-*myc* DNA-binding protein. Mol Cell Biol 1985; 5:448-56.
32. Leatherwood J, Lopez-Girona A, Russell P. Interaction of Cdc2 and Cdc18 with a fission yeast ORC2-like protein. Nature 1996; 378.
33. Millar JBA, Russell P, Dixon JE et al. Negative regulation of mitosis by two functionally overlapping PTPases in fission yeast. EMBO J 1992; 11:4943-52.
34. Millar JBA, Buck V, Wilkinson MG. Pyp1 and Pyp2 PTPases dephosphorylate an osmosensing MAP kinase controlling cell size at division in fission yeast. Genes Dev 1995; 9:2117-30.

Chapter 10

Expression and Secretion of Mammalian Proteins in *Schizosaccharomyces pombe*

Kumar Sambamurti

Introduction

Saccharomyces cerevisiae is a widely used host for the expression of eukaryotic proteins. However, due to problems such as poor recognition of foreign signal sequences, improper processing and poor secretion, alternative eukaryotic expression systems such as *Pichia pastoris* have been recently developed.[1] The fission yeast *Schizosaccharomyces pombe* resembles higher eukaryotes in several aspects such as transcriptional regulation, splicing and glycosylation and may therefore serve as a good model system for expression of proteins from higher eukaryotes.[2] Several mammalian promoters including SV40[3] and CMV[4] are expressed well in this yeast and some mammalian introns are spliced.[5] In addition, *S. pombe* has secreted glycoproteins with with terminal galactose[6,7] residues, not found in other yeasts. Furthermore, unlike *S. cerevisiae*, indigenous secretory proteins in *S. pombe* such as acid phosphatase are known to be secreted into the culture medium[8-10] which simplifies production and analysis of the secreted protein.

S. pombe has been successfully used to clone several mammalian proteins by functional complementation.[11,12] Despite these advantages, this yeast has not been studied very extensively, mutants defining the secretory pathway have not been isolated and secretion of mammalian proteins have not been studied in detail. However, *S.*

Foreign Gene Expression in Fission Yeast Schizosaccharomyces pombe, edited by Yuko Giga-Hama and Hiromichi Kumagai.

pombe mutants deficient in galactosylation have been recently isolated reflecting renewed interest in this organism.[13,14] A few mammalian proteins have been shown to be secreted from *S. pombe;* they include mouse α-amylase,[15] and anti-thrombin-III.[16]

Here we present a preliminary report on expression of a secreted derivative of the human placental alkaline phosphatase[17] (SEAP) in *S. pombe* under a regulated promoter fructose bisphosphatase[18] (fbp; described elsewhere in this book). Placental alkaline phosphatase (PLAP) is widely used as a tumor marker and is known to be a membrane glycoprotein anchored via a phosphatidyl inositol glycan (PIG) tail.[19] The SEAP protein is a truncated derivative of PLAP and has the advantage that it can be readily assayed in the medium by simple colorimetric and chemiluminescent methods.[17] The protein can tolerate both N-terminal and C-terminal fusion without loss of activity and it has been shown to be a glycoprotein with N-linked sugars in mammalian cells.[17,20-22] The SEAP protein can be potentially exploited for analysis of gene regulation, secretion, protein topology, protein turnover and membrane anchoring. Several elegant genetic studies in bacterial secretion used bacterial alkaline phosphatase similarly.[23] This study tests the feasibility of using SEAP as a reporter in *S. pombe* with the long term goal of optimizing the *S. pombe* expression system and studying membrane protein processing.

Plasmid Construction, Transformation and Assay

Materials, Cell Cultures and Plasmids

Strains and plasmids used are summarized in Table 10.1. The yeast strain AZ16, the inducible vector pCHY21 and the SEAP cDNA were generous gifts from Dr. Holly Prentice, Dr. Robert E. Kingston, Dr. Charles Hoffman and Dr. Brian Cullen respectively. The SEAP cDNA was subcloned as a *Sal*1-Xho1 fragment into the Xho1 site of pCHY21 using standard techniques[24] and transformed into AZ16 cells

Table 10.1 Strains and plasmids

AZ16 *S. pombe*	h^- leu1, ura4	Holly Prentice and Robert Kingston[26]
pCHY21	$Ura3^+$, FBP promoter	Charles Hoffman[18]
pBC12/CMV/SEAP	SEAP	Brian Cullen[17]
pCHY/SEAP	pCHY21 + SEAP	This work

by the lithium acetate method.[25] Media (MB and YE) were prepared as described,[2] supplemented with CSM-ura-leu (Bio101) + 150 mg/ml leucine and either 3% dextrose (dex) or 3% glycerol + 0.1% dextrose (gly) for regulated expression of the fbp promoter.[18] Fresh overnight cultures in MB-dex were diluted into MB-gly or YE-gly to about 2 x 10^5 cells per ml, cultured overnight to log phase (5 x 10^6 cells/ml) and an additional 8 hr to stationary phase (2 x 10^7 cells/ml) before use for further analysis. MB-gly media (5 ml) from SEAP-secreting cells were adjusted to pH 7.5 with sodium phosphate and loaded over 1 ml columns of Concanavalin A agarose (ConA; Sigma) or biotinylated *Bandeiraea simplicifolia* lectin 1 (BSL1; Pierce, immobilized on avidin-agarose column) and the unbound material was assayed for alkaline phosphatase activity.

Cell Lysates and Western Blot Analysis

Cell suspensions (50 ml) were centrifuged at 2000 rpm for 10 min, washed once in distilled water at 4°C and resuspended at 2.5 x 10^9 cells/ml in homogenization buffer (0.1 M Tris pH 8, 1 mM DTT, 20% glycerol). Approximately 4 g/ml of acid-washed glass beads (0.5 mm; Sigma) were added to the suspension and cells were lysed by vortexing for 5 x 1 min at 4°C. Lysates were recovered after washing beads once with an equal volume of buffer. Twenty five microliters corresponding to 3 x 10^7 cells were separated on 10-20% Tris-tricine polyacrylamide gels[27] (Novex) for Western blot analysis by the enhanced chemiluminescence method. Briefly, the protein was transferred onto immobilon P membranes (Millipore) in 0.2 M Tris, 0.15 M glycine containing 20% methanol for 2 hr on ice. The transferred protein was blocked in 3% BSA overnight, incubated with anti-PLAP antibodies (1:2000 DAKO) for 2 hr, washed 6 times in TBST (0.1 M Tris, 0.15 M NaCl pH 8 containing 0.1% Tween 20), treated with secondary antibody (donkey anti-rabbit-peroxidase conjugated, 1:2000; Amersham) for 1 hr and washed 6 times in TBST. The washed blots were rinsed in TBS, treated with chemiluminescence substrate (Amersham) for 1 min, and exposed to autoradiography film.

Alkaline Phosphatase Assays

Alkaline phosphatase activity in the medium was detected by spotting culture media (0.8 ml) on nitrocellulose filters in a dot blotter and staining the filter with Nitroblue tetrazolium/ Bromo-chloro-indolyl phosphate (1-step NBT/BCIP; Pierce) or by assaying individual filter circles using para-nitro-phenyl phosphate (PNPP;

Kirkegaard and Perry Laboratories, Inc.) as a soluble substrate and reading the sample at 405 nm in a microtiter plate in diethanolamine buffer as suggested by the manufacturer.

Results

Expression of SEAP Protein in Cell Lysates

Figure 10.1 is a Western blot analysis of SEAP expression in cell lysates from untransformed controls (lanes 1 and 3) and cells transformed with pCHY/SEAP (lanes 2 and 4) and grown in either MB-gly (lanes 1 and 2) or YE-gly (lanes 3 and 4). The levels of the SEAP protein appear to be much higher in lysates from MB-gly than YE-gly grown cells. Although detectable levels of SEAP expression are seen in these lysates, most of the protein is secreted into the medium (data not shown). Western blots of lysates prepared in the presence of a protease inhibitor cocktail (Boehringer Mannheim) did not show any increase in SEAP level (data not shown), suggesting that the cellular pool of SEAP protein is not subject to extensive proteolysis under conditions used for cellular lysis.

Characterization of the Secreted SEAP Protein

Secretion of the SEAP protein was monitored by enzyme activity in the medium as described in materials and methods. The data are presented in Figures 10.2 and 10.3 and in Table 10.2. The reference numbers A1-H2 are kept constant in both figures for easy cross reference. Substantial levels of alkaline phosphatase activity were seen in both MB-gly and YE-gly media in SEAP-transformed cul-

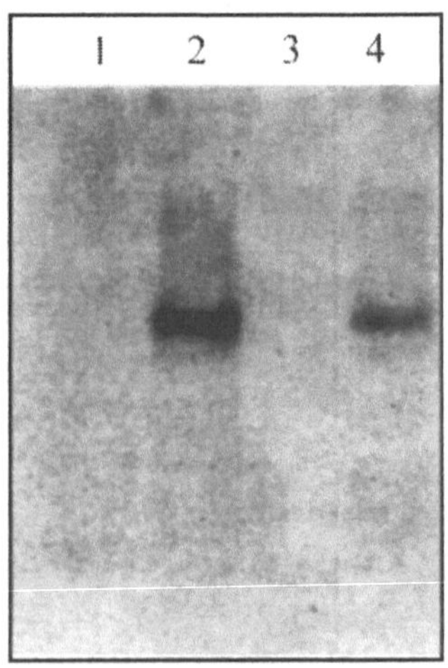

Fig. 10.1. Characterization of SEAP in *S. pombe* cell-lysates. Cell lysates from control (lanes 1 and 3) *S. pombe* cultures and those transformed with pCHY/SEAP (lanes 2 and 4) were prepared from cells grown in MB-gly (lanes 1 and 2) and YE-gly (lanes 3 and 4) media and subjected to Western blot analysis as described in materials and methods. Note that a band of ~55 kDa is detected in transformed and not in control cell lysates.

tures (Fig. 10.2, 3-D1, F1) but not in controls (Fig. 10.2, 3-C1, E1). The protein is lost from the medium when loaded over either a ConA or a BSL1 column, suggesting that the protein is glycosylated with both mannose and galactose residues (Fig. 10.2, 3-H1,H2). The recombinant SEAP protein is stable upon heating to 60°C for 15 min (Fig. 10.2, 3-G1) like its mammalian counterpart.[17]

Similar data showing SEAP secretion into culture media along with heat stability and lectin binding were obtained by a quantitative enzyme assay using the soluble PNPP substrate (Table 10.2 and Fig.10. 3). Note that higher levels of SEAP activity were detected in YE-gly media (Table 10.2/Fig. 10.3-F1) than in MB-gly media (Table 10.2/Fig. 10.3-D1). Low levels of SEAP activity are also detected in MB-dex (Table 10.2/Fig. 10.3-B1) probably due to sugar depletion during growth as previously described.[18]

Regulation of SEAP Activity

The SEAP protein was expressed under a regulated promoter in order to test its value as a reporter, and expression of SEAP activity in the medium was monitored in both the log and stationary phases under non-inducing and inducing conditions. The data obtained show that levels of alkaline phosphatase activity are very low in MB-dex (Fig. 10.2, 3-B1) due to repression of the fbp promoter by dextrose as previously described.[18] Growth in glycerol results in derepression of the fbp promoter with 20 and 30 fold induction of SEAP activity in MB-gly and YE-gly respectively (Fig. 10.2, 3-D1, E1).

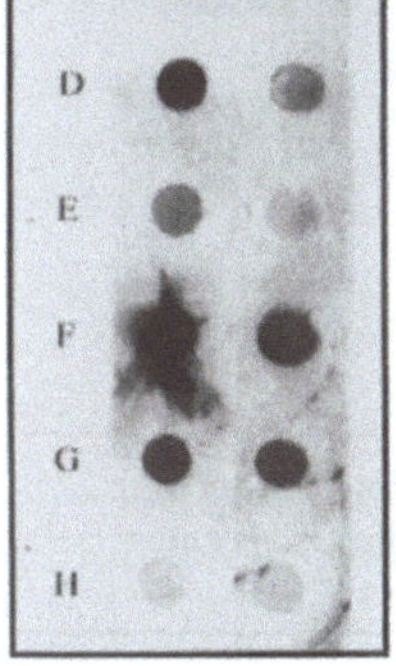

Fig. 10.2. Detection of SEAP secreted into *S. pombe* culture media. Nitrocellulose filters were spotted with 0.8 ml of the following log phase (A1-F1) and stationary phase (A2-F2) *S. pombe* media: control (A, C and E); SEAP-transformed (B, D, F) and stained as described in materials and methods. Sample G1 was subjected to 60°C for 15 min and G2 was adjusted to pH 7.5 prior to assaying for activity and samples H1 and H2 were first adsorbed over a ConA and BSL1 column respectively.

Steady state levels of SEAP activity in media were high in log phase cultures (Fig. 10.2, 3-D1, E1) and low in the stationary phase (Fig. 10.2, 3-D2, E2), suggesting that in addition to poor expression in the stationary phase, the protein is degraded rapidly in the medium. Expression was higher in YE-gly media, suggesting that either the protein was expressed, albeit at low levels, in the stationary phase or that the secreted protein was more stable in YE-gly. The latter is suggested by the finding that SEAP levels in YE-gly medium were higher in log phase cultures in spite of lower levels in cell lysates (Fig. 10.1).

Discussion

The data show that SEAP can be readily expressed and secreted in *S. pombe* using its own signal sequence. It is interesting to note that the signal sequence of SEAP works well in *S. pombe* in spite of the presence of a stretch of seven tandem rare leucine codons (CTG).[30]

The protein is glycosylated using mannose as well as galactose residues as shown by binding of ConA and BSL1. Levels of secreted protein are somewhat higher in YE-gly media than in MB-gly media. High levels of expression are seen in log phase cultures but levels of secreted SEAP activity drop to zero in the stationary phase minimal media, indicating that the protein is not expressed in stationary phase medium and that the accumulated secreted protein is degraded drop in SEAP activity was also observed in YE-gly but sub-

Table 10.2. **Expression of alkaline phosphatase in S. pombe culture media**

			1 OD at 405nm 2	
			Log phase	Stationary
I	Untransfected control	A	0.024	0.012
MB-Dex	SEAP	B	0.071	0.034
II	Untransfected control	C	0.015	0.082
MB-Gly	SEAP	D	0.971	0.089
III	Untransfected control	E	0.023	0.02
YE-Gly	SEAP	F	1.408	0.508
	Heated SEAP	G1	0.255	
	ConA SEAP	H1	0.005	
	pH SEAP	H1	0.246	
	BS1 Lectin	H2	0.011	
	Pure PLAP 200 ng		0.849	

Media (0.8 ml/well) were captured on nitrocellulose membranes in a dot blotter. Dots were punched out and assayed for alkaline phosphatase activity as described in materials and methods. Letters A-H correspond to labels in Fig. 10.2.

stantial activity remained, suggesting that the secreted protein was either more stable in this medium or synthesis continued at a lower level. Others have shown that robust α-amylase expression required the presence of protease inhibitors such as chymostatin[15] which unfortunately could not be used as it inhibited SEAP activity (data not shown). Two proteases responsible for modulation of the mating response have been described and further studies are needed to reveal their role in degrading secreted proteins.[29]

The results indicate that *S. pombe* is a strong candidate for the expression of mammalian proteins but that it needs to be fine tuned by optimization of vector systems[28] and isolation of protease deficient and/or hyper-secreting mutants.

The SEAP protein can be expressed and regulation of gene expression monitored by directly assaying media without having to prepare cell lysates. The secreted proteins from colonies can be trapped on nitrocellulose filters and detected either by alkaline phosphatase activity or by immunostaining (Fig. 10.4). This provides a simple assay for isolating mutants and clones that affect the secretory pathway. Stability of SEAP in the medium under various conditions is currently being studied and mutants that improve expression of SEAP are being isolated to characterize the secretory and proteolytic pathways in this important model organism.

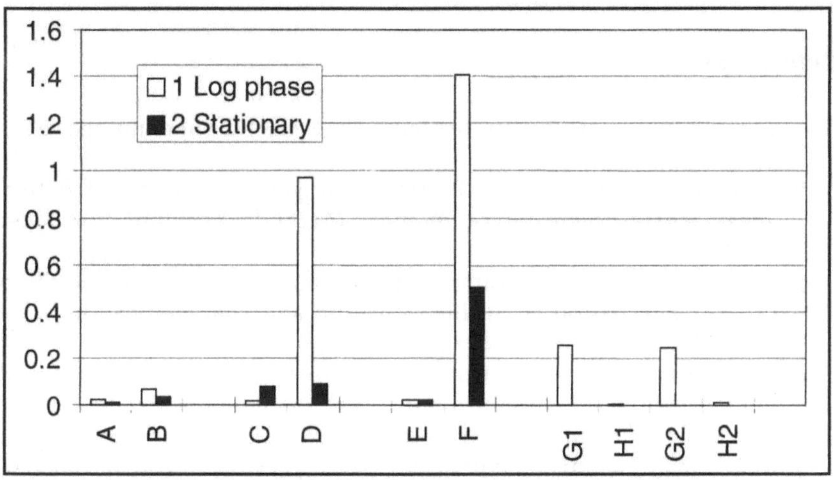

Fig. 10.3. Secreted SEAP activity determined in *S. pombe* culture media. Alkaline phosphatase activity was measured using PNPP as a substrate as described in materials and methods. Individual bar graphs are labeled as in Figure 10.2.

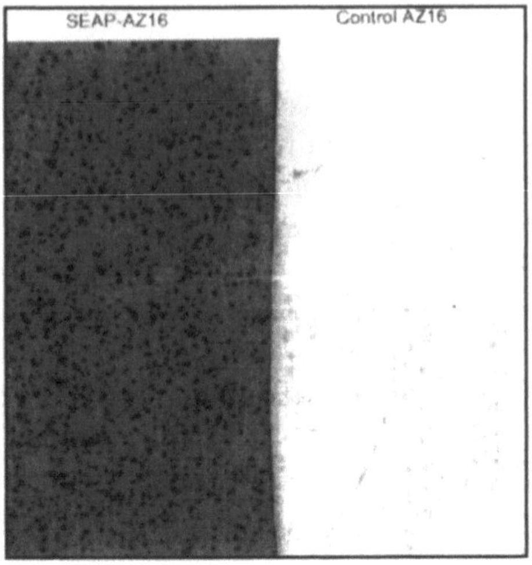

Fig. 10.4. Colony screen for SEAP secretion in transformed *S. pombe*. S. pombe (AZ16) colonies transformed with SEAP and controls were cultured on nitrocellulose filters in MB-gly + agar dishes. The colonies were washed out and filters immunostained for SEAP secretion. Note the dark spots of SEAP indicating the position of colonies in transformed cells alone.

Acknowledgments

The author would like to thank Dr. Charles Hoffman, Dr. Brian Cullen, Dr. Holly Prentice, and Dr. Robert Kingston for generous gifts of plasmids and *S. pombe* strains which made this work possible and Dr. Michael Cherry for permission to use the codon usage tables deposited by him on the Internet. The author would also like to thank Dr. NK Robakis and Dr. MA Pappolla for laboratory support during early stages of this study and Dr. SG Younkin for his support and encouragement at the Mayo Clinic, where the bulk of the data presented was generated. Finally, the author would like to thank Dr. Kiminobu Sugayo for help with the illustrations.

References

1. Buckholz RG, Gleeson MA. Yeast systems for the commercial production of heterologous proteins. [Review] Biotechnology 1991; 9(11):1067-72.
2. Moreno S, Klar A, Nurse P. Molecular genetic analysis of fission yeast *Schizosaccharomyces pombe*. Methods Enzymol 1991; 194: 795-823.
3. Jones RH, Moreno S, Nurse P, Jones NC. Expression of the SV40 promoter in fission yeast: Identification and characterization of an AP-1-like factor. Cell 1988; 53(4):659-67.
4. Toyama R, Okayama H. Human chorionic gonadotropin alpha and human cytomegalovirus promoters are extremely active in the fission yeast *Schizosaccharomyces pombe*. FEBS Lett 1990; 268(1):217-21.
5. Kaufer NF, Simanis V, Nurse P. Fission yeast *Schizosaccharomyces pombe* correctly excises a mammalian RNA transcript intervening sequence. Nature 1985; 318(6041):78-80.

6. Chappell TG, Warren G. A galactosyltransferase from the fission yeast *Schizosaccharomyces pombe.* J Cell Biol 1989; 109(6 Pt 1):2693-702.
7. Sanchez Y, Moreno S, Rodriguez L., Synthesis of *Saccharomyces cerevisiae* invertase by *Schizosaccharomyces pombe.* FEBS Lett 1988; 234(1):95-9.
8. Mitchison JM, Creanor J. Linear synthesis of sucrase and phosphatases during the cell cycle of *Schizosaccharomyces pombe.* J Cell Sci 1969; 5(2):373-91.
9. Maundrell K, Nurse P, Schonholzer F. et al. Cloning and characterization of two genes restoring acid phosphatase activity in pho1-mutants of *Schizosaccharomyces pombe.* Gene 1985; 39(2-3):223-30.
10. Elliott S, Chang CW, Schweingruber ME et al. Isolation and characterization of the structural gene for secreted acid phosphatase from *Schizosaccharomyces pombe.* J Biol Chem 1986; 261(6):2936-41.
11. Lee MG, Nurse P. Complementation used to clone a human homologue of the fission yeast cell cycle control gene cdc2. Nature 327(6117):31-5, 1987 May 7-13.
12. Xu HP, Wang Y, Riggs M et al. Biological activity of the mammalian RAP genes in yeast. Cell Regul 1990; 1(10):763-9.
13. Ballou L, Ballou C. *Schizosaccharomyces pombe* mutants that are defective in glycoprotein galactosylation. Proc Natl Acad Sci USA 1995; 92(7):2790-4.
14. Huang KM, Snider MD. Isolation of protein glycosylation mutants in the fission yeast *Schizosaccharomyces pombe.* Mol Biol Cell 1995; 6(5):485-96.
15. Tokunaga M, Kawamura A, Yonekyu S et al. Secretion of mouse alpha-amylase from fission yeast *Schizosaccharomyces pombe:* presence of chymostatin-sensitive protease activity in the culture medium. Yeast 1993; 9(4):379-87.
16. Bröker M, Ragg H, Karges HE. Expression of human antithrombin III in *Saccharomyces cerevisiae* and *Schizosaccharomyces pombe.* Biochem Biophys Acta 1987; 908(3):203-13.
17. Berger J, Hauber J, Hauber R et al. Secreted placental alkaline phosphatase: a powerful new quantitative indicator of gene expression in eukaryotic cells. Gene 1988; 66(1):1-10.
18. Hoffman CS, Winston F. Glucose repression of transcription of the *Schizosaccharomyces pombe* fbp1 gene occurs by a cAMP signaling pathway. Genes Dev 1991; 5(4):561-71.
19. Cross GA. Eukaryotic protein modification and membrane attachment via phosphatidylinositol. [Review] Cell 1987; 48(2):179-81.
20. Flanagan JG, Leder P. The kit ligand: A cell surface molecule altered in steel mutant fibroblasts. Cell 1990; 63(1):185-94.
21. Efthimiopoulos S, Felsenstein KM, Sambamurti K et al. Study of the phorbol ester effect on Alzheimer amyloid precursor processing: sequence requirements and involvement of a cholera toxin sensitive protein. J Neurosci Res 1994; 38(1):81-90.

22. Pappolla MA, Sambamurti K, Efthimiopoulos S et al. Heat-shock induces abnormalities in the cellular distribution of amyloid precursor protein (APP) and APP fusion proteins. Neurosci Lett 1995; 192(2):105-8.
23. Ehrmann M, Boyd D, Beckwith J. Genetic analysis of membrane protein topology by a sandwich gene fusion approach. Proc Natl Acad Sci USA 1990; 87(19):7574-8.
24. Sambrook J, Fritsch EF, Maniatis T. Molecular Cloning. Cris Nolan, ed. Cold Spring Harbor Laboratory Press, 1989.
25. Okazaki K, Okazaki N, Kume K et al. High-frequency transformation method and library transducing vectors for cloning mammalian cDNAs by trans-complementation of *Schizosaccharomyces pombe*. Nucleic Acids Res 1990; 18(22):6485-9.
26. Gallo GJ, Prentice H, Kingston RE. Heat shock factor is required for growth at normal temperatures in the fission yeast *Schizosaccharomyces pombe*. Mol Cell Biol 1993; 13(2):749-61.
27. Schagger H, von Jagow G. Tricine-sodium dodecyl sulfate-polyacrylamide gel electrophoresis for the separation of proteins in the range from 1 to 100 kDa. Anal Biochem 1987; 166(2):368-79.
28. Giga-Hama Y, Tohda H, Okada H, et al. High-level expression of human lipocortin I in the fission yeast *Schizosaccharomyces pombe* using a novel expression vector. Biotechnology (N Y) 1994; 12(4):400-4.
29. Imai Y, Yamamoto M. *Schizosaccharomyces pombe sxa1+* and *sxa2+* encode putativeproteases involved in the mating response. Mol/Cell Biol 1992; 12(4):1827-34.
30. Cherry M Codon usage table on the internet: (cherry@frodo.mgh.harvard.edu).

CHAPTER 11

Secretion of Human Interleukin-6 Using the P-Factor Secretion Signal in *Schizosaccharomyces pombe*

Yuko Giga-Hama

Introduction

The vectors for heterologous gene expression in *S. pombe* have been developed and high-level productions of several proteins have been achieved. In the case of the secretory proteins, however, few proteins have successfully been secreted. Secretory proteins must enter the secretory pathway (composed mainly of endoplasmic reticulum (ER) and Golgi apparatus accompanied with the removal of the signal sequence of each protein by a signal peptidase in the host cells for correct folding, disulfide bonds, glycosylation, etc. Some heterologous secretory proteins such as human antithrombin III,[1] human gastric lipase,[2] human placental alkaline phosphatase (chapter 10) and *S. cerevisiae* invertase[3,4] can be produced extracellulary with the removal of their own signal sequence, while some signal sequences of secretory proteins are not recognized by the signal peptidase in *S. pombe*. The signal sequences functionable in *S. pombe* have not yet been exploited extensively except for the secretion signal of the killer toxin of *Kluiveromyces lactis*.[5] The development of convenient secretion vectors is important for the production of secretory proteins with a natural structure.

Foreign Gene Expression in Fission Yeast Schizosaccharomyces pombe, edited by Yuko Giga-Hama and Hiromichi Kumagai.

Imai and Yamamoto recently cloned a gene of the P-factor precursor protein, Map2, which is one of the *S. pombe* mating pheromones.[6] After several processing steps, P-factor is thought to be secreted into the culture medium, like the α-mating factor of *S. cerevisiae*.[7] We suspect that the signal region of Map2 may be available and effective for heterologous protein secretion. However, the detailed processing mechanism of P-factor has not been elucidated, and its signal sequence has not yet been identified.

In this chapter, we will demonstrate that human interleukin-6, one of the cytokines, was secreted effectively into the culture medium using the Map2 signal region, and we will show that a possible pre-pro region of the P-factor precursor was determined.

Human Interleukin-6

Human interleukin-6 (hIL-6), originally identified as a B cell-stimulatory factor-2 (BSF-2)[8] or interferon β-2,[9] is a major cytokine which has been shown to play a central role in the physiology of several tissues and organs. It is produced by macrophages, T-and B-lymphocytes, fibroblasts and other types of cells. It regulates immune responses in acute phase reactions and hematopoiesis.[10] It also induces an increase in platelets, suggesting that IL-6 may be useful in the treatment of decreased platelet production due to various causes.[11] hIL-6 is a secretory glycoprotein composed of 184 amino acids; its apparent molecular weight is around 21-28 kDa. The poduction of functionally active hIL-6 is important for structural/functional studies, and medical and pharmaceutical purposes. hIL-6 was expressed intracellulary in *E. coli* without a signal peptide[12-14] and was shown to be secreted by *Aspergillus nidulans*.[15] hIL-6 was also successfully secreted by *S. cerevisiae* using the secretory signal of the α-mating factor.[16] In *S. pombe* we have not yet achieved the secretion of hIL-6 in the medium using its own signal sequence, but we found its precursor accumulated in the cytosol of *S. pombe*. We concluded that for hIL-6 secretion it is necessary to use a suitable signal sequence derived from another protein.

P-factor

P-factor is a diffusible mating pheromone secreted by *S. pombe* h^+ cells.[6] *S. pombe* has two mating types, called h^+ and h^-. h^+ and h^- mate with each other and form diploid zygotes when they are starved for nutrition. Both h^+ and h^- secrete pheromones named P-factor and M-factor,[17] respectively. M-factor is a nonapeptide, the carboxy-terminal cysteine residue of which is both s-farnesylated and carboxymethylated.[18] It is not secreted through the ordinary secretory pathway, and it has no "signal sequence."

P-factor is a peptide of 23 residues with the sequence Thr-Tyr-Ala-Asp-Phe-Leu-Arg-Ala-Tyr-Gln-Ser-Trp-Asn-Thr-Phe-Val-Asn-Pro-Asp-Arg-Pro-Asn-Leu. This peptide is produced by the processing of the precursor protein which is encoded by the *map2* gene. Imai and Yamamoto[6] reported that the predicted *map2* gene product (Map2) has 201 amino acid residues and that its amino terminus is hydrophobic, suggesting that this region might serve as a signal sequence. Map2 carries four tandem copies of 34 amino acid residues, each of which terminated with tribasic residues Lys-Lys-Arg (Fig. 11.1a). Two possible N-linked glycosylation sites were seen near the carboxyl terminus of Map2, unlike that of the α-factor precursor protein. In these four tandem copies, the second and the fourth copies (hatched boxes in Figure 11.1a) have exactly the same sequence, but the sequence of the other copies are slightly different. The sequence of purified active P-factor shows that it is the same as that of the second and fourth regions. It is not certain whether the putative peptides derived from the first and the third region in the four tandem copies have P-factor activity.

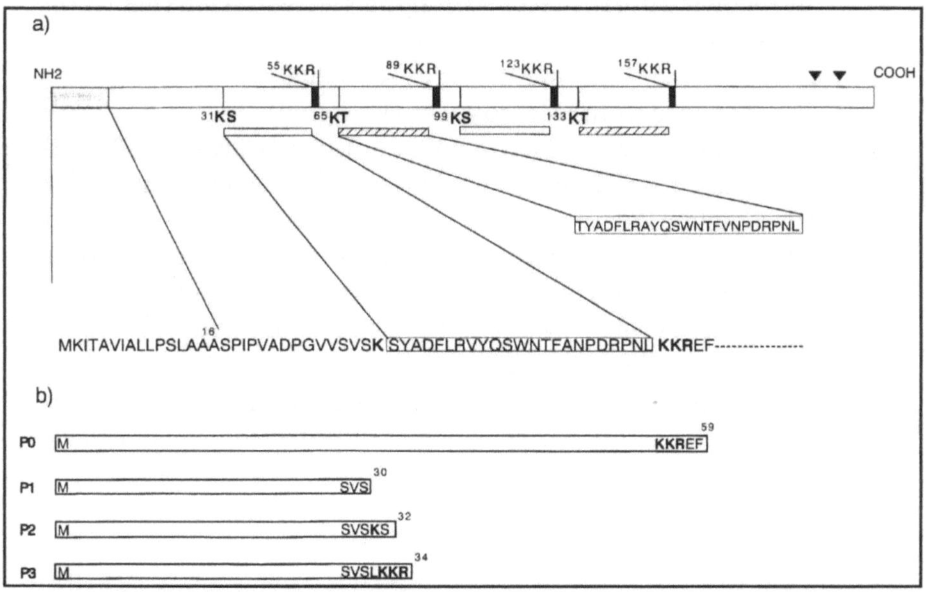

Fig. 11.1. (a.) The structure of the P-factor precursor protein (Map2).[6] Both the white boxes and hatched boxes are P-factors (the two hatched boxes have the same sequence and were found to be active). The shadowed box is the N-terminal, the putative "pre" sequence. Each amino acid is abbreviated by one letter. Black triangles: possible N-glycosylation sites. (b.) Putative secretory signals ; Po, P1, P2, P3.

Although the secretory signal of Map2 is considered to be a possible candidate of the signal for foreign protein secretion, the precise signal region and the detailed processing mechanism have not been elucidated.

Examination of the Secretory Signal of Map2

By an analogy of the processing mechanism of the α-mating factor in *S. cerevisiae* P-factor is probably processed from a precursor protein (Map2) by several proteases such as signal peptidase, Krp1,[19] carboxy peptidase, aminopeptidase and some other processing enzymes. Judging from the specificity of signal peptidase[20] and other processing enzymes, together with the feature of the primary structure of Map2 and purified P-factor, several processing sites are predicted as follows:

1) The carboxy terminus of Lys-Lys-Arg sequences;
2) Between Lys and Ser or the Lys and Thr sequence found at amino acid numbers 31-32, 65-66, 99-100, 133-134;
3) Between 16Ala and 17Ser, 18Pro and 19Ile, Ala 22 and Asp 23, Pro 24 and Gly 25 (these are candidates for a putative signal peptidase recognition site according to the report of von Heijne[20]). Based on this speculation regarding processing sites, the following putative secretory signals were selected (Fig.11.1b):

1) The sequence from 1Met to 59Phe (designated as P0 signal);
2) The sequence from 1Met to 30Ser (P1 signal);
3) The sequence from 1Met to 32Ser (P2 signal);
4) The sequence from 1Met to 30Ser + Leu-Lys-Lys-Arg (P3 signal). These four kinds of signals were followed by an hIL-6 (or hIL-6 mutant) mature region, and each fused gene was inserted into pTL2M1 (chapter 4) to express hIL-6 extracellularly. After the cultivation of *S. pombe* transformants, the medium containing secreted proteins was analyzed by sodium dodecyl sulfate-polyacrylamide gel electrophoresis (SDS-PAGE) followed by Western blot analysis. The N-terminal amino acid sequence analysis of protein bands which were detected as hIL-6 clarified the actual processing sites.

Secretion of hIL-6

Cloning of hIL-6 cDNA

hIL-6 cDNA was cloned from a cDNA library of human peripheral blood monocyte (LPS-activated; Clontech Laboratory Inc, Palp

Alto, CA) according to the oligonucleotide probe method. The DNA probe was synthesized based on the hIL-6 sequence determined by Hirano et al[8] using DNA synthesizer (Model 381A, Applied Biosystem Inc, CITY COUNTRY). The synthesized N-terminal and C-terminal probes were 5'-CAA CAC CAG GAG CAG CCC CAG GGA GAA GGC AAC TGG ACC GAA GGC GCT-3' (antisense of nucleotides 52-99) and 5'-ATG ACA ACT CAT CTC ATT CTG CGC AGC TTT AAG GAG TTC CTG CAG TCC-3' (nucleotides 598-645), respectively. cDNA was cloned by the plaque hybridization method according to the attached manual (Clontech) using a cDNA library prepared with λgt11 as a cloning vector and C600 as a host *E. coli*. The hIL-6 cDNA insert was isolated from λgt11 by digestion with *Eco*RI and ligated to pUC19 which was already digested with *Eco*RI, giving a new plasmid named pAG-9-8-1.

Construction of the Secretion Vector Using Secretory Signal Po

The open reading frame (ORF) of mature hIL-6 was amplified by polymerase chain reaction (PCR) using two primers (5'-GAA TTC ATG CCA GTA CCC CCA GGA GAA GAT-3', 5'- AAG CTT ATT ACA TTT GCC GAA GAG CCC TCA G-3') and pAG-9-8-1 as a template. PCR products were digested with both *Eco*RI and *Hin*dIII and subjected to agarose gel electrophoresis. An approximately 600 bp fragment was cut out and purified according to the glass powder method (designated as fragment A).

Map2 secretion signal Po (see above) was amplified by PCR using two primers (5'- CGG TCA TGA AGA TCA CCG CTG TCA T-3', 5'-GGG AAG CTT AGC TCT CAA ATT TGG CAG-3') and pADMP2[6] which contains the full length of *map2* as a template. The PCR products were digested with *Bsp*HI and *Eco*RI and subjected to acrylamide gel electrophoresis. An approximately 200 bp fragment was cut out and eluted from the gel, giving a signal fragment (designated as fragment B).

Expression vector pTL2M1 was digested with both *Afl*III and *Hin*dIII and was subjected to agarose gel electrophoresis. The band corresponding to about 5,000 bp was cut out and purified with the glass powder method (designated as fragment C).

The three fragments (A, B, and C) were ligated with T4 DNA ligase. *E. coli (DH5)* was transformed with this ligated product. *E. coli* containing the correctly constructed vector was screened. The resulting vector, pSL2P06m (Fig. 11.2.a), which has mature IL-6 (185

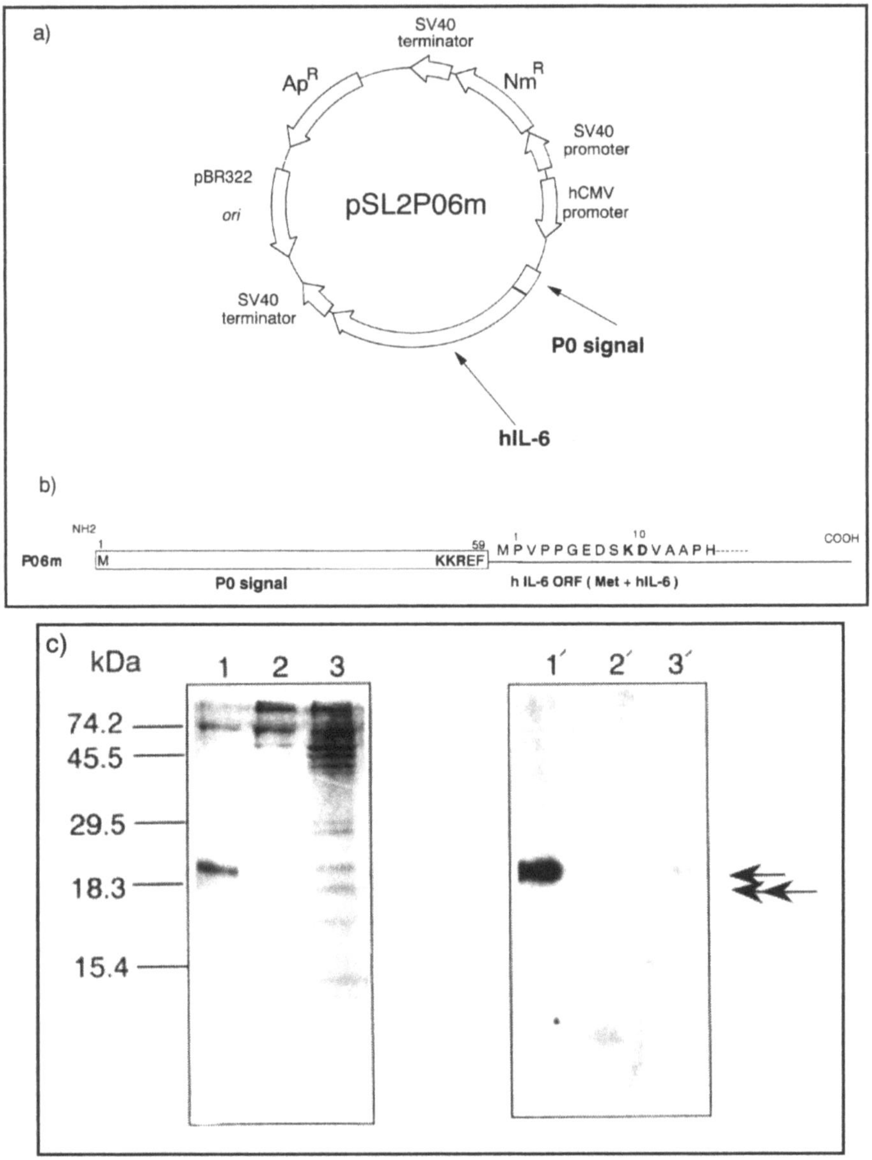

Fig. 11.2. a) Secretion vector for human IL-6 using the Po secretory signal. hCMV, human cytomegalovirus; SV40, simian virus; Nm^R, neomycin-resistance-encoding gene; Ap^R, ampicillin-resistance-encoding gene; *ori*, replication origin. (b.) The structure of the fusion protein (Po6m) of both Po secretory signal and mature hIL-6. (c.) SDS-PAGE (left) and Western blot analysis using anti-hIL-6 antibody (right) of culture medium of *S. pombe*/pSL2Po6m. Lanes 1 and 1', 1 µg of recombinant hIL-6 (*E. coli*). Lanes 2 and 2', concentrated culture medium of *S. pombe*/pTL2M1 (no insert, control). Lane 3 and 3', concentrated culture medium of *S. pombe*/pSL2Po6m. Single-head and double-head arrows indicate EF+M+hIL-6 and its degradation product, respectively. For the culture conditions and sample preparation, see "Expression and secretion of hIL-6", below.

residues) fused to Po signal (59 residues) was confirmed by restriction enzyme digestion and partial DNA sequence analysis. The N-terminal structure of the fused protein Po6m is schematically shown in Figure 11.2.b.

Expression and Secretion of hIL-6

S. pombe (*leu1-32h⁻* strain) was transformed by both pSL2Po6m and the transducing vector pAL7 according to the method of Okazaki et al,[21] and transformants were screened by G418.[22] Transformed *S. pombe* was cultivated in MM medium[23] containing 1% casamino acid, 2% glucose and G418 (25 μg/mL) at 32°C for 24 hr. The cultivated *S. pombe* was inoculated into MM medium containing 1% casamino acid, 2% glucose and G418 (200 μg/mL) at 32°C for 18 hr, and then centrifuged. The supernatant was concentrated by 100-fold with a membrane filter (Amicon, Beverly, MA). The concentrated medium was analyzed by SDS-PAGE, followed by Western blot analysis (Fig. 11.2.c). Two clear bands around 20 kDa were detected by Western blot analysis.

The N-terminal amino acid sequences of these two bands, which were transferred to a PVDF membrane, were analyzed by a gas phase protein sequencer (Shimadzu PSQ-1, Kyoto, Japan). The N-terminal

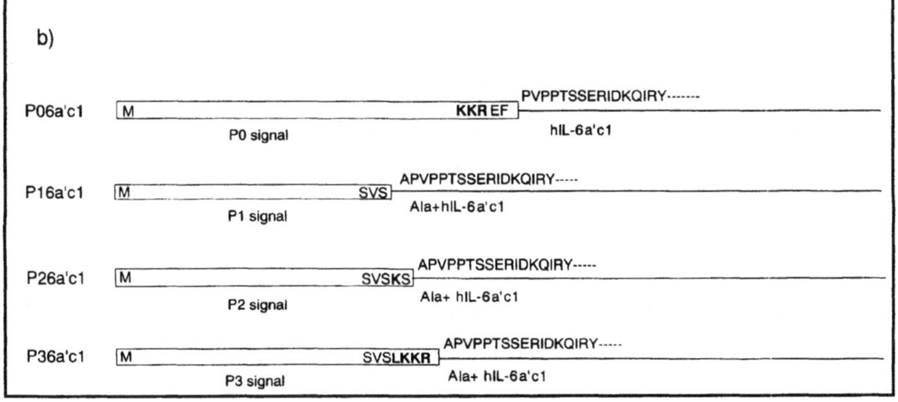

Fig. 11.3. (a.) The structure of the human IL-6 mutant, hIL6a'c1. Black boxes are deleted. (b.) The structure of the fusion proteins of secretory signals (Po, P1, P2, P3) and hIL-6a'c1. These fusion proteins were designated as Po6a'c1, P16a'c1, P26a'c1, P36a'c1, respectively.

amino acid sequences of the upper band and lower bands are Glu-Phe-Met-Pro-Val-Pro-Pro- and Asp-Val-Ala-Ala-Pro-His-, respectively. From the results of the upper band, hIL-6 was found to be correctly processed between Arg 57 and Glu 58 in Po signal during the secretion pathway and to be secreted into the medium. The N-terminal amino acid sequence of lower band showed that an unexpected proteolysis between Lys 9 and Asp 10 of hIL-6 (Fig. 11.2c) occurred. To exclude this proteolysis, a deletion mutant of hIL-6, hIL-6a'c1, was used for the subsequent analyses.

Secretion of hIL-6 mutant hIL-6a'c1

Characteristics of hIL-6a'c1

Many deletion mutants were constructed, and the relationship between their activity and structure has been discussed.[24,25] We have also constructed many mutants from which were deleted the N-terminal region and/or the region containing disulfide-bonded Cys. Each mutant gene was expressed using the expression vector pTL2M1 (chapter 4). For the analysis of the secretion signal, hIL-6a'c1 (163 amino acids) which has two deleted regions (amino acids 5-19 and 44-50 of hIL-6, Fig. 11.3.a) was chosen because it has no Lys 9-Asp 10 and because the produced protein shows good activity using the hIL-6a'c1 expression vector designated pTL26a'c1(unpublished data). The apparent molecular weight of hIL-6a'c1 on SDS-PAGE is around 18 kDa.

Construction of Secretion Vector for hIL-6a'c1 Using Secretory Signal Po

The ORF of hIL-6a'c1 was amplified by PCR using two primers (5'-AAA GAA TTC CCA GTA CCC CCA ACC TCT TCA-3', 5'-AAG CTT ATT ACA TTT GCC GAA GAG CCC TCA G-3') and pTL26a'c1 as a template. The resulting fragment was digested with both *Eco*RI and *Hin*dIII, and subjected to agarose gel electrophoresis. An approximately 600 bp fragment was cut out and purified according to the glass powder method. The recombinant vector pSL2Po6m (subsection "Construction of the secretion vector using secretory signal PO") was also digested with both *Eco*RI and *Hin*dIII, and a fragment of around 5,000 bp was purified by the glass powder method. These two fragments were ligated with T4 DNA ligase. *E. coli* (*DH5*) was transformed with this ligation product. The resulting vector, pSL2Po6a'c1, which has the gene of hIL-6a'c1 (163 residues) fused to the Po signal (59 residues) was screened using *E. coli* (*DH5*) and confirmed by restriction enzyme digestion and partial DNA sequence analysis. The N-terminal structure of the fused protein Po6a'c1 is schematically shown in Figure 11.3.b.

Expression and Secretion of hIL-6a'c1 Using Secretory Signal Po

S. pombe (*leu1-32h*$^-$ strain) was transformed by both pSL2Po6a'c1 and the transducing vector pAL7 according to the method of Okazaki et al,[21] and transformants were screened by G418.[22] Transformed *S. pombe* was cultivated in MM medium[23] containing 1% casamino acid, 2% glucose, and G418 (25 µg/mL) at 32°C for 24 hr. The cultivated *S. pombe* was inoculated into MM medium containing 1% casamino acid, 2% glucose, and G418 (200 µg/mL) at 32°C for 24 hr, and then centrifuged. The supernatant was concentrated by 100-fold with a membrane filter (Amicon). The concentrated medium was analyzed by SDS-PAGE, followed by Western blot analysis (Fig. 11.4). An approximately 18 kDa single band was detected by Western blot analysis.

The N-terminal amino acid sequence of this single band, which was transferred to a PVDF membrane, was analyzed by the gas phase protein sequencer (Shimadzu). The N-terminal amino acid sequence

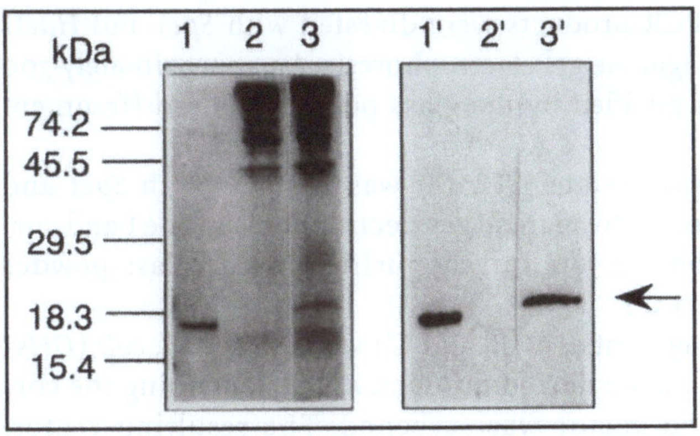

Fig. 11.4. SDS-PAGE (left) and Western blot analysis using anti-hIL-6 antibody (right) of culture medium of *S. pombe*/pSL2Po6a'c1. Lanes 1 and 1', 1 µg of recombinant hIL-6a'c1 (*E. coli*). Lanes 2 and 2', concentrated culture medium of *S. pombe*/pTL2M1 (no insert, control). Lane 3 and 3', concentrated culture medium of *S. pombe*/pSL2Po6a'c1. Arrow shows EF+ hIL-6a'c1. For culture conditions and sample preparation, see "Expression and secretion of hIL-6a'c1 using secretory signal PO".

of this band was Glu-Phe-Pro-Val-Pro-Pro-. No other bands were detected by Western blot analysis. These results demonstrated that hIL-6a'c1 was secreted as a single protein by the processing of the Po signal, and showd that it is not necessary to cleave the linkage between Lys 9 and Asp 10 for the secretion. Although other processing must have occurred in the upstream region of the Po signal in the secretion pathway, the cleavage between Arg 57 and Glu 58 was important and was compled as a final step of the processing. hIL-6a'c1 is thus the proper molecule to elucidate the secretion signal. To elucidate the other processing sites, the following experiment was performed.

Construction of the Secretion Vector for hIL-6a'c1 Using Secretory Signal P1

The ORF of mature hIL-6a'c1 was amplified by PCR using two primers (5'-CTT GGC GCC CCA GTA CCC CCA ACC TCT TC-3', 5'-AAA ATG ATT TAA AGG CTA TA-3') and pSL2Po6a'c1 as a template. The resulting fragment was digested with *Hae*II and *Hin*dIII and subjected to agarose gel electrophoresis. An approximately 500 bp fragment was cut out and purified using the glass powder method (fragment A').

The promoter and secretion signal region was also synthesized by PCR using two primers (5'-TTG ACT AGT TAT TAA TAG TA-3', 5'-CCA AGC GCT AAC TGA AAC CAC ACC AG-3') and pSL2Po6a'c1 as a template. The PCR products were digested with *Spe*I and *Hae*II and subjected to agarose gel electrophoresis. An approximately 700 bp fragment was purified by the glass powder method (fragment B').

The expression vector pTL2M1 was digested with *Spe*I and *Hin*dIII and subjected to agarose gel electrophoresis. The band corresponding to about 4,500 bp was purified by the glass powder method (Fragment C').

The three fragments, (A', B' and C') were ligated. *E. coli* (*DH5*) was transformed by the ligated product. *E. coli* containing the correctly constructed vector was screened. The resulting vector, pSL2P16a'c1, which has mature hIL-6a'c1 (163 residues) fused to the P1 signal (30 residues) was confirmed by restriction enzyme digestion and partial DNA sequence analysis. The N-terminal structure of the fused protein P16a'c1 is schematically shown in Figure 11.3.b.

Expression and Secretion of hIL-6a'c1 Using Secretory Signal P1

S. pombe (*leu1-32h*⁻ strain) was transformed by both pSL2P16a'c1 and the transducing vector pAL7 according to the method of Okazaki et al,[21] and transformants were screened by G418.[22] Transformed *S. pombe* was cultivated in MM medium[23] containing 1% casamino acid, 2% glucose, G418 (25 μg/mL) at 32°C for 24 hr. The cultivated *S. pombe* was inoculated into MM medium containing 1% casamino acid, 2% glucose, and G418 (200 μg/mL) at 32°C for 24 hr, and then centrifuged. The supernatant was concentrated by 100-fold with a membrane filter (Amicon). The concentrated medium was analyzed by SDS-PAGE, followed by Western blot analysis (Fig.11.5, lanes 3 and 3'). Two bands around 19 kDa were detected by Western blot analysis.

The N-terminal amino acid sequences of these two bands, which were transferred to a PVDF membrane, were analyzed by the gas phase protein sequencer. The N-terminal amino acid sequence of the upper and lower bands are Asp-Pro-Gly-Val-Val-Ser-Val-Ser-Ala-Pro- and Gly-Val-Val-Ser-Val-Ser-Ala-Pro-Val-Pro-, respectively. Based on the results of the upper band, hIL-6a'c1 was found to be processed between Ala 22 and Asp 23 and also between Pro 24 and Gly 25 of the secretion signal (Fig. 11.1a) during the secretion pathway, and was shown to be secreted into the medium. It is not clear whether the cleavage between 24Pro and 25Gly was caused by a signal peptidase or by diamino peptidase after the cleavage by signal peptidase.

Construction of the Secretion Vector for hIL-6a'c1 Using Secretory Signal P2

The ORF of mature hIL-6a'c1 was amplified by PCR using two primers (5'-CTT GGC GCC CCA GTA CCC CCA ACC TCT TC-3', 5'-AAA ATG ATT TAA AGG CTA TA-3') and pSL2P06a'c1 as a template. The resulting fragment was digested with *Hae*II and *Hin*dIII and subjected to agarose gel electrophoresis. An approximately 500 bp fragment was cut out and purified using the glass powder method (fragment A").

The promoter and secretion signal region was also synthesized by PCR using two primers (5'-TTG ACT AGT TAT TAA TAG TA-3', 5'-CCA AGC GCT CTT GCT AAC TGA AAC CAC AC-3') and pSL2P06a'c1 as a template. The PCR products were digested with *Spe*I and *Hae*II and subjected to agarose gel electrophoresis. An approximately 700 bp fragment was purified by the glass powder method (fragment B").

The expression vector pTL2M1 was digested with *Spe*I and *Hin*dIII and subjected to agarose gel electrophoresis. The band corresponding to about 4,500 bp purified by the glass powder method (Fragment C").

The three fragments, (A", B" and C") were ligated. The *E. coli* (*DH5*) was transformed by this ligated product. The *E. coli* containing correctly constructed vector was screened. The resulting vector, pSL2P26a'c1, which has mature IL-6a'c1 (163 residues) fused to the P2 signal (32 residues) was confirmed by restriction enzyme digestion and partial DNA sequence analysis. The N-terminal structure of the fused protein P26a'c1 is schematically shown in Figure 11.3.b.

Expression and Secretion of hIL-6a'c1 Using Secretory Signal P2

S. pombe (*leu1-32h*$^-$ strain) was transformed by both pSL2P26a'c1 and the transducing vector pAL7 according to the method of Okazaki et al,[21] and transformants were screened by G418.[22] Transformed *S. pombe* was cultivated in MM medium[23] containing 1% casamino acid, 2% glucose, and G418 (25 µg/mL) at 32°C for 24 hr. The cultivated *S. pombe* was inoculated into MM medium containing 1% casamino

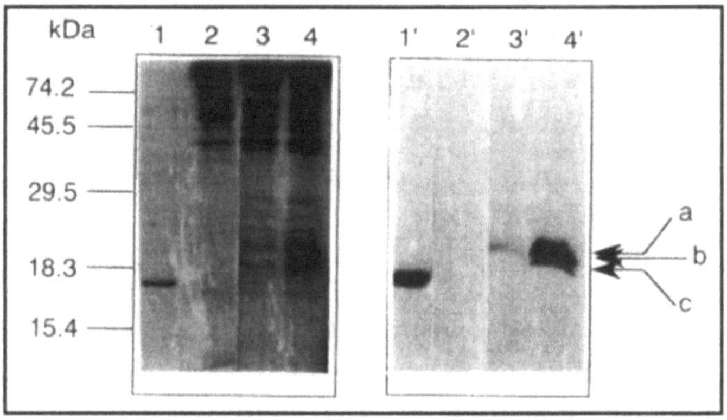

Fig. 11.5. SDS-PAGE (left) and Western blot analysis using anti-hIL-6 antibody (right) of culture medium of *S. pombe*/pSL2P16a'c1 and *S. pombe*/pSL2P26a'c1. Lanes 1 and 1', 1 µg of recombinant hIL-6a'c1 (*E. coli*). Lanes 2 and 2', concentrated culture medium of *S. pombe*/ pTL2M1 (no insert, control). Lanes 3 and 3', concentrated culture medium of *S. pombe*/pSL2P16a'c1. Lane 4 and 4', concentrated culture medium of *S. pombe*/pSL2P26a'c1. Arrows (a), (b) and (c) show DPGVVSVSKS+A+hIL-6a'c1, GVVSVSKS+A+hIL-6a'c1 and S+A+hIL-6a'c1, respectively. For culture conditions and sample preparation, see "Expression and secretion of hIL-6a'c1 using secretory signal P1" and "Expression and secretion of hIL-6a'c1 using secretory signal P2".

acid, 2% glucose, and G418 (200 μg/mL) at 32°C for 24 hr, and then centrifuged. The supernatant was concentrated by 100-fold with a membrane filter (Amicon). The concentrated medium was analyzed by SDS-PAGE, followed by Western blot analysis (Fig. 11.5 lanes 4 and 4'). Two bands of about 19 kDa and one band around 18.5 kDa were detected by Western blot analysis.

The N-terminal amino acid sequence of these three bands, which were transferred to PVDF membrane, was analyzed by the gas phase protein sequencer. The N-terminal amino acid sequences of the upper band, lower band and 18.5 kDa band are Asp-Pro-Gly-Val-Val-Ser-Val-Ser-Ala-Pro-, Gly-Val-Val-Ser-Val-Ser-Ala-Pro-Val-Pro- and Ser-Ala-Pro-Val-Pro-Pro-Thr-Ser-Ser-Glu, respectively. Based on these results, hIL-6a'c1 was found to be processed between Lys 31 and Ser 32 in addition to Ala22 and Asp 23, Pro 24 and Gly 25, of the secretion signal during the secretion pathway, and was shown to be secreted into the medium.

Construction of the Secretion Vector for hIL-6a'c1 Using Secretory Signal P3

The ORF of mature hIL-6a'c1 was synthesized by PCR using two primers (5'-TTT CTT AAG AAG CGT CCA GTA CCC CCA ACC TCT TC-3', 5'-AAA ATG ATT TAA AGG CTA TA-3') and pSL2P06a'c1

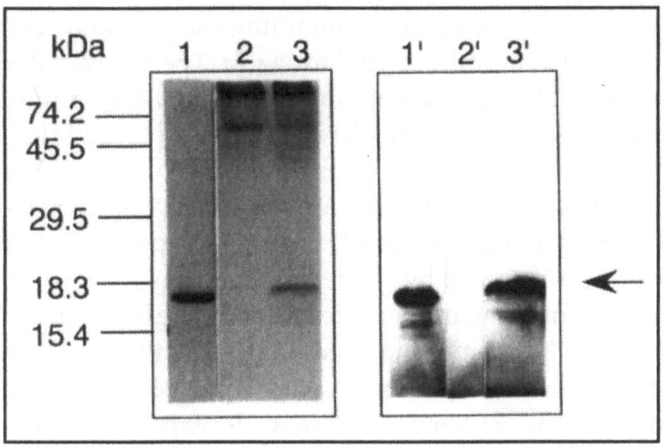

Fig. 11.6. SDS-PAGE (left) and Western blot analysis using anti-hIL-6 antibody (right) of culture medium of *S. pombe*/pSL2P36a'c1. Lane1 and 1', 1 μg of recombinant hIL-6a'c1 (*E. coli*). Lane 2 and 2', concentrated culture medium of *S. pombe*/pTL2M1 (no insert, control). Lane 3 and 3', concentrated culture medium of *S. pombe*/pSL2P36a'c1. Arrow shows A+hIL-6a'c1. For culture conditions and sample preparation, see subsection "Expression and secetion of hIL-6a'c1 using secretor signal P3."

as a template. The PCR products were digested with *Hae*II + *Hin*dIII, and the approximately 500 bp fragment was purified by the glass powder method (fragment A'''). The promoter and secretion signal region was also synthesized by PCR using two primers (5'-TTG ACT AGT TAT TAA TAG TA-3', 5'-CTT CTT AAG GCT AAC TGA AAC CAC ACC AG-3') and pSL2P06a'c1 as a template. The PCR products were digested with *Spe*I + *Afl*II, and the approximately 700 bp fragment was purified by glass powder method (fragment B''').

The expression vector pTL2M1 was digested with *Spe*I and *Hin*dIII, and the approximately 4,500 bp fragment was purified by the glass powder method (Fragment C''').

The three fragments, (A''', B''' and C''') were ligated and screened using *E. coli* (*DH5*). The resulting vector, pSL2P36a'c1. which has mature IL-6a'c, (163 residues) fused to the P3 signal (34 residues) was confirmed by restriction enzyme mapping and partial DNA sequence analysis. The N-terminal structure of the fused protein P36a'c1 is shown schematically in Figure 11.3.b.

Expression and Secretion of hIL-6a'c1 Using Secretory Signal P3

S. pombe (*leu1-32h*$^-$ strain) was transformed by both pSL2P36a'c1 and the transducing vector pAL7 according to the method of Okazaki et al,[21] and transformants were screened by G418.[22] Transformed *S. pombe* was cultivated in MM medium[23] containing 1% casamino acid, 2% glucose, and G418 (25 µg/mL) at 32°C for 24 hr. The cultivated *S. pombe* was inoculated into 2 x MM medium containing 2% casamino acid, 10% glucose, and G418 (400 µg/mL) at 32°C for 24 hr, and then centrifuged. The supernatant was concentrated by 10-fold with a membrane filter. The concentrated medium was analyzed by SDS-PAGE, followed by Western blot analysis (Fig. 11.6). An approximately 18 kDa single band was detected by Western blot analysis.

The N-terminal amino acid sequence of this single band, which was transferred to a PVDF membrane, was analyzed by the gas phase protein sequencer. The N-terminal amino acid sequence of this band was Ala-Pro-Val-Pro-Pro-Thr-Ser-Ser-Glu-Arg-Ile-Asp-Lys-Gln-Ile-Arg-. Based on this result, hIL-6a'c1 was found to be correctly processed at the carboxy terminal of Lys-Lys-Arg of the P3 signal.

Conclusion

Although it is important to enter the secretion pathway in order to make the correct structure similar to or the same as that of the natural protein, not many secretion signals available in *S. pombe* has been reported thus far.[5] In the present investigation, hIL-6 and its mutants were efficiently secreted into the medium by novel secretion vectors with the P-factor secretion signal. Although P-factor is known to be one of the secreted peptides and plays an important role in mating of both h^- and h^+ *S. pombe*, its processing mechanism from P-factor precursor protein has not been elucidated, and the secretion signal available for foreign protein secretion remains obscure.

We selected four putative secretion signals, P0, P1, P2 and P3, and constructed secretion vectors for hIL-6 or hIL-6a'c1. Secreted hIL-6 was detected as two bands on SDS-PAGE using the P0 signal. However, one band was produced by over-processing between Lys 9 and Asp 10 in hIL-6. In the case of hIL-6a'c1 with the P0 signal, a single band was detected on SDS-PAGE, suggesting that the cleavage between Lys 9 and Asp 10 is not necessary for the protein secretion. The P0 signal was removed correctly between Arg 57 and Glu 58, and hIL6a'c1 was efficiently secreted into the medium.

Based on the N-terminal sequences of each protein secreted in the medium, the cleavage sites were elucidated, as shown in Figure 11.7. The precise processing mechanism of the P-factor precursor which occurs in the secretion pathway is not yet clear. However judging from these data and the prediction by Heijine,[20] the sequence from Met 1 to Ala 22 may be a "pre" sequence followed by the "pro" sequence from Asp 23 to Lys 31. At this stage, it is not clear whether the carboxyterminal of Pro 24 is cleaved by signal peptidase or by another processing enzyme such as dipeptidyl aminopeptidase. The cleavage of the carboxyl end of Lys 31 was thought to have been done by the trypsin-like processing protease in the ER or Golgi apparatus. The sequence from Ser 32 to Leu 54 probably corresponds to one of the four P-factors, although it is not known whether it is active (see section "P-factor").

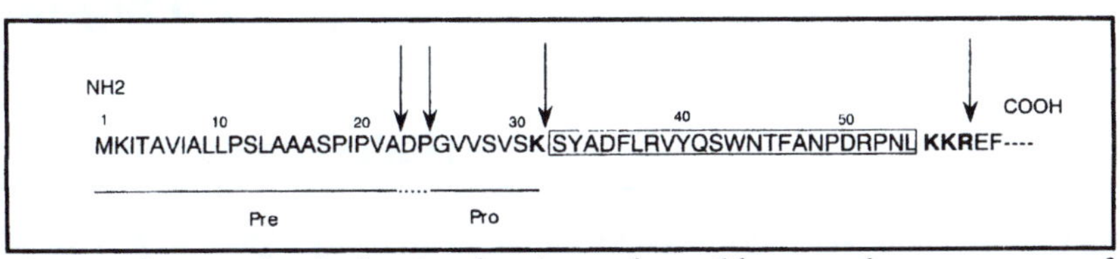

Fig. 11.7. Cleavage sites in the signal region and possible pre and pro sequences of P-factor precursor protein. Arrows show the cleavage sites determined from this experiment.

Judging from the results of the secretion of the P1 and P2 signals fused to hIL-6a'c1, the long sequence corresponding to the putative P-factor as shown in the P0 signal is not always necessary for protein secretion. However, the Lys-Lys-Arg sequence is important to obtain homogeneous target protein, because in the case of the P1 and P2 signals, the processing was not completely done and several components were secreted. Actually, the P3 signal which has no P-factor sequence was successfully removed and hIL-6a'c1 was efficiently secreted as a single protein. This result also suggests that the Glu-Phe sequence after Lys-Lys-Arg is not always necessary for correct processing.

Based on these data, we propose the P3 signal as the secretion signal and the secretion vector pSL2P3M4 (Fig. 11.8) for foreign secretory protein production.

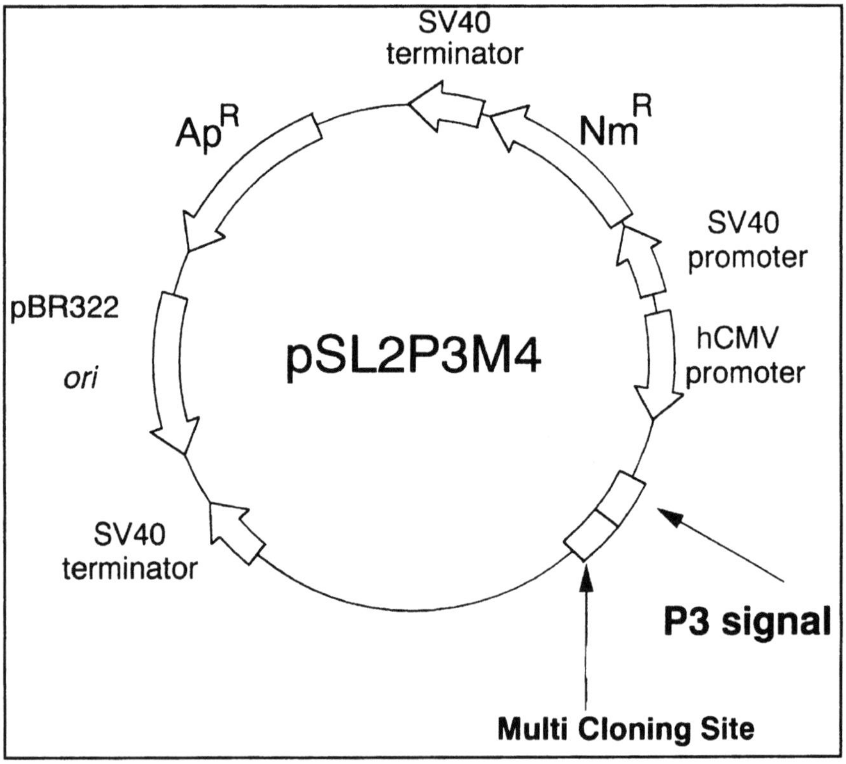

Fig. 11.8. Secretion vector pSL2P3M4 with the P3 secretory signal. hCMV, human cytomegalovirus; SV40, simian virus; Nm^R, neomycin-resistance-encoding gene; Ap^R, ampicillin-resistance-encoding gene; *ori*, replication origin.

Future Perspectives

For more efficient secretion of heterologous proteins using *S. pombe* as a host, several aspects such as the prevention of unexpected proteolysis, improvement of the host, optimal culture conditions, large-scale fermentation using a jar fermenter and other factors are important, in addition to the secretion signal.

Regading the prevention of the unnecessary cleavage of secreted proteins, Tokunaga reported that the addition of protease inhibitor into the culture medium was effective.[5] Culture conditions such as the concentration of each composition of medium, culture time, temperature, etc. are also important for the unexpected proteolysis. The starvation of carbon and/or nitrogen sources caused an increase in the protease expression.[26] Protease-deficient mutants are also considered to be a promising host for the prevention of unexpected proteolysis. The protease B-like protein or carboxypeptidase Y-deficient mutants which were obtained by Sato[26] and Takegawa[27], respectively, are possible candidates. Foreign protein secretion using these protease-deficient mutants as hosts is now under investigation.

The screening of super-secretion mutants is effective for obtaining large amounts of the secreted protein. We obtained super-secreted *S. pombe* strain in a preliminary experiment by the screening of mutants produced by the alkylating reagent ethyl methanesulfonate (EMS). These mutants were detected by the Western blot analysis of each colony. We can now use one of the super-secretion *S. pombe* strains which shows 10 times higher secretion efficiency than that of the original *S. pombe* (unpublished). Further screenings are under investigation. In the future, many *S. pombe* mutants with improved properties such as super secretion, efficient growth, protease deficiency etc. will be obtained by the screening of mutants and/or direct mutations of target genes of *S. pombe*. The elucidation of the entire sequence of *S. pombe* chromosome DNA by the *S. pombe* genome analysis project (see chapter 1) is expected to provide ways to make some "convenient" mutants more easily.

To obtain large amounts of cell mass, it is necessary to examine the culture conditions.[28,29] In the case of the secretion of hIL-6/hIL-6a'c1 (subsection "Expression and secretion of hIL-6a'c1, using secretory signal PO" and "Expression and secretion of hIL-6a'c1 using secretory signal P3"), the final cell density in the medium of 2 x MM with 2% casamino acid and 10% glucose was several times higher than that in the MM medium with 1% casamino acid and 2%

glucose by flask culture. Along with the increase of the cell density, the amount of secreted protein is increased, as shown in Figures 11.4 and 11.6. We are continuing to investigate culture conditions using a jar fermenter for industrial applications. The development of fermentation technology usingS. *pombe* is of course important to obtain high amounts of products per liter culture at low cost. Several approaches have been tried thus far.[28,29]

Secretory protein production is becoming increasingly useful and informative not only for basic research but also for industrial applications. There are limits to the production of complex structured proteins using prokaryotic microorganisms as a host. We are now trying to produce many kinds of secretory proteins and to develop this production system for industrial use.

Acknowledgments

I am grateful to Drs. Masayuki Yamamoto and Yoshiyuki Imai, the University of Tokyo, for their gift of pADMP2. I would like to thank Dr. Masafumi Yohda, the Institute of Physical and Chemical Research (RIKEN), and Robert Ingham, University of British Columbia, for their support during the early stages of this study. I would also like to thank Mr. Hideki Tohda, Ms. Hiroko Tsukamoto, Mr. Kiyokazu Nikaido and Dr. Hiromichi Kumagai, Asahi Glass Co, Ltd, for their support and stimulating discussions. I thank Dr. Atsushi Isoai, Asahi Glass Co, Ltd, for his continuous encouragement.

References

1. Bröker M, Ragg H, Karges HE. Expression of human antithrombin III in *Saccharomyces cerevisiae* and *Schizosaccharomyces pombe*. Biochem Biophys Acta 1987; 908:203-213.
2. Smerdon GR, Aves SJ, Walton EF. Production of human gastric lipase in the fission yeast *Schizosaccharomyces pombe*. Gene 1995; 165:313-318.
3. Zárate V, Belda F. Characterization of the heterologous invertase produced by *Schizosaccharomyces pombe* from the SUC2 gene of *Saccharomyces cerevisiae*. J Appl Bacteriol 1996; 80: 45-52.
4. Sánchez Y, Moreno S, Rodríguez L. Synthesis of *Saccharomyces cerevisiae* invertase by *Schizosaccharomyces pombe*. FEBS Lett 1988; 234:95-99.
5. Tokunaga M, Kawamura A, Yonekyu S et al. Secretion of mouse α-amylase on fission yeast *Schizosaccharomyces pombe*. Yeast 1993; 9:379-387.
6. Imai Y, Yamamoto Y. The fission yeast mating pheromone P-factor: Its molecular structure, gene structure, and ability to induce

gene expression and G1 arrest in the mating partner. Genes & Dev 1994; 8:328-338.
7. Julius D, Schekman R, Thorner J. Glycosylation and processing of prepro-α-factor through the yeast secretory pathway. Cell 1984; 36:309-318.
8. Hirano T, Yasukawa K, Harada H et al. Complementary DNA for a novel human interleukin (BSF-2) that induces B-lymphocytes to produce immunoglobulin. Nature 1986; 324:73-76.
9. Zilberstein A, Ruggieri R, Korn JH et al. Structure and expression of cDNA and genes for human interferon β-2, a distinct species inducible by growth-stimulatory cytokines. EMBO 1986; 5:2529-2537.
10. Lotz M. Interleukin-6. Cancer Invest 1993; 11:732-742.
11. Ishibashi T, Kimura H, Shikama Y. Interleukin-6 is a potent thrombopoietic factor in vivo in mice. Blood 1989; 74:1241-1244.
12. Tonouchi N, Oouchi N, Kashima N et al. High-level expression of human BSF-2/IL-6 cDNA in *Escherichia coli* using a new type of expression-preparation system. J Biochem 1988; 104:30-34.
13. Yasueda H, Nagase K, Hosoda A et al. High-level direct expression of semi-synthetic human interleukin-6 in *Escherichia coli* and production of N-terminus Met-free product. Bio/Technology 1990; 8:1036-1040.
14. Arcone R, Pucci P, Zappacosta F et al. Single-step purification and structural characterization of human interleukin-6 produced in *Escherichia coli* from a T7 RNA polymerase expression vector. Eur J Biochem 1991; 198:541-547.
15. Carrez D, Janssens W, Degrave P et al. Heterologous gene expression by filamentous fungi: secretion of human interleukin-6 by *Aspergillus nidulans.* Gene 1990; 94:147-154.
16. Guisez Y, Tison B, Vandekerckhove J et al. Production and purification of recombinant human interleukin-6 secreted by the yeast *Saccharomyces cerevisiae.* Eur J Biochem 1991; 198:217-222.
17. Davy J. Isolation and quantitation of M-factor, a diffusible mating factor from the fission yeast *Schizosaccharomyces pombe.* Yeast 1991; 7:357-366.
18. Davy J. Mating pheromones of the fission yeast *Schizosaccharomyces pombe:* Purification and structural characterization of M-factor and isolation and analysis of two genes encoding the pheromone. EMBO J 1992; 11:951-960.
19. Davy J, Davis K, Imai Y et al. Isolation and characterization of krp, a dibasic endopeptidase required for cell viability in the fission yeast *Schizosaccharomyces pombe.* EMBO 1994; 13:5910-5920.
20. von Heijne G. A new method for predicting signal sequence cleavage sites. Nuc Acid Res 1986; 14:4683-4690.
21. Okazaki K, Okazaki N, Kume K et al. High-frequency transformation method and library transducing vectors for cloning mammalian cDNAs by trans-complementation of *Schizosaccharomyces pombe.* Nucleic Acid Res 1990; 18:6485-6489.

22. Tohda H, Okada H, Giga-Hama Y, Okayama H, and Kumagai H. A copy-number-controlled expression vector for the fission yeast *Schizosaccharomyces pombe.* Gene 1994; 150:275-280.
23. Alfa C, Fantes P, Hyams J et al. *S. pombe* growth media. In: Experiments with Fission Yeast: Cold Spring Harbor Laboratory Press, Cold Spring Harbor, New York. 1993:133-136.
24. Brakenhoff JPJ, Hart M, Groot ERD et al. Structure-function analysis of human IL-6. J Immunol 1990; 145:561-568.
25. Snouwaert JN, Leebeek FWG, Fowlkes DM et al. Role of disulfide bonds in biologic activity of human interleukin-6. J Biol Chem 1991; 266:23097-23102.
26. Satoh S, Suzuki H, Widyastuti U. Identification and characterization of genes induced during sexual differentiation in *Schizosaccharomyces pombe.* Current Genetics 1994; 26:31-37.
27. Tabuchi M, Iwaihara O, Ohtani Y et al. Vacuolar protein sortin in fission yest: cloning, biosynthesis, transport and cprocessing of carboxypeptidase Y from *Schizosacharomyces pombe.* J Bacteriol 1997; 179(13): in press.
28. Hde Queiroz J, Uribelarrea J-L, Pareilleux A. High cell density cultures of *Schizosaccharomyces pombe* in a cell-recycle reactor. Appl Biochem and Biotech. 1991; 30:285-295.
29. Hde Queiroz J, Pareilleux A. Growth kinetics of *Schizosaccharomyces pombe* under various culture conditions: influence of pH, maltose, ethanol and oxygenation. Appl Microbiol Biotechnol 1990; 33:578-581.

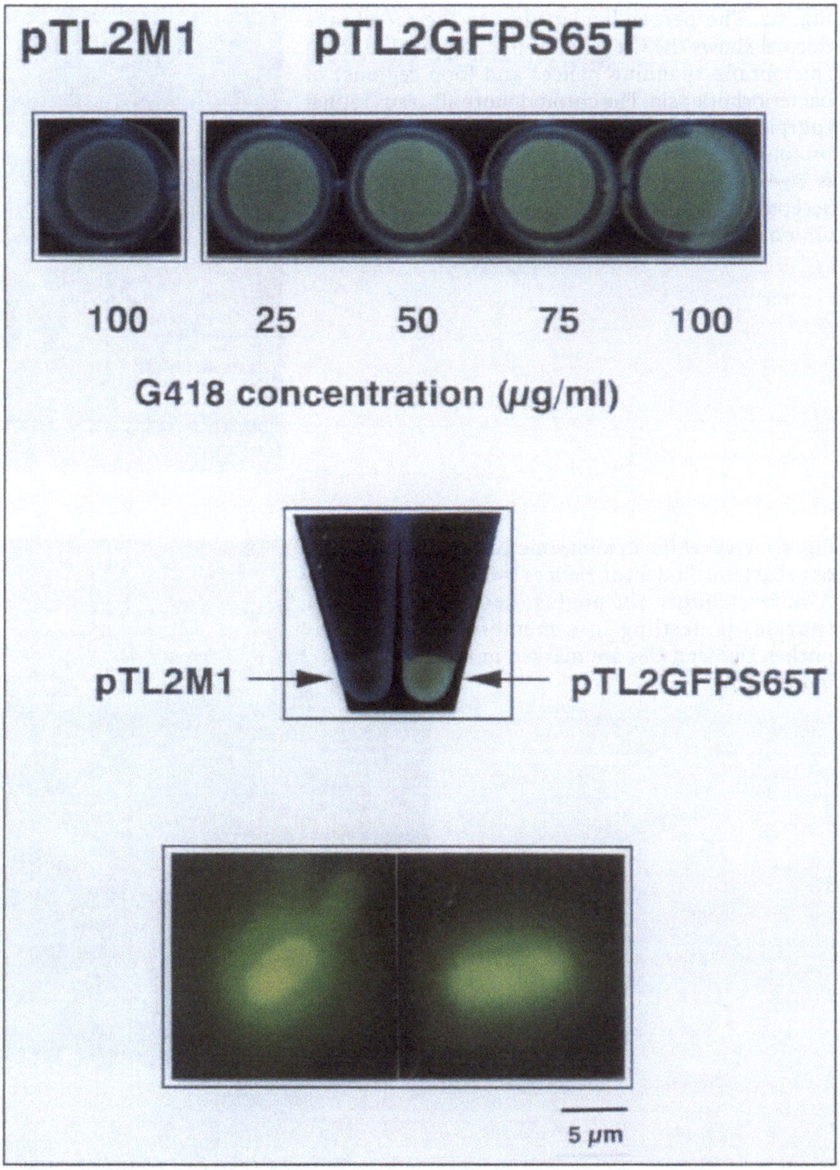

Fig. 4.4b. Detection of the fluorescence of GFP expressed in recombinant yeast. Transformants in a culture plate (upper panel) and harvested cells (middle panel) showed green fluorescence under long-wave UV light. Each cell also fluoresces (lower panel) under fluorescent microscopy (Olympus, Tokyo, Japan).

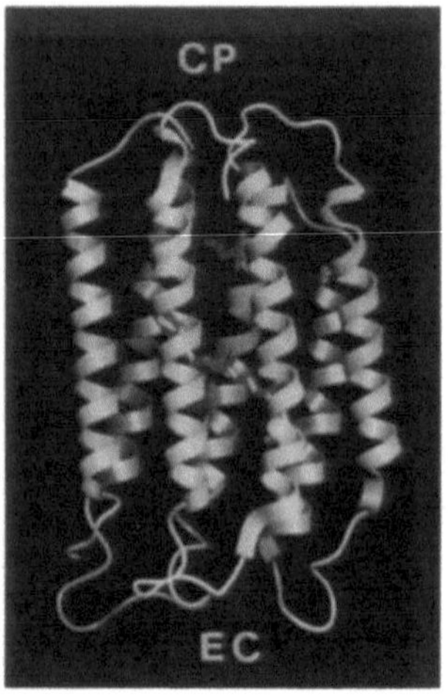

Fig. 5.2. The perpendicular view to the membrane normal shows the Cα trace of the polypeptide chain (membrane-spanning helices and loop regions) of bacteriorhodopsin. The chromophore all-trans retinal (purple) is covalently bound to K216 in helix G. The proton donor for the light induced proton translocation is D96 in the cytoplasmic channel (CP). The proton acceptor is D85 in the extracellular channel (EC). Both amino acid residues (red) are located in helix C.

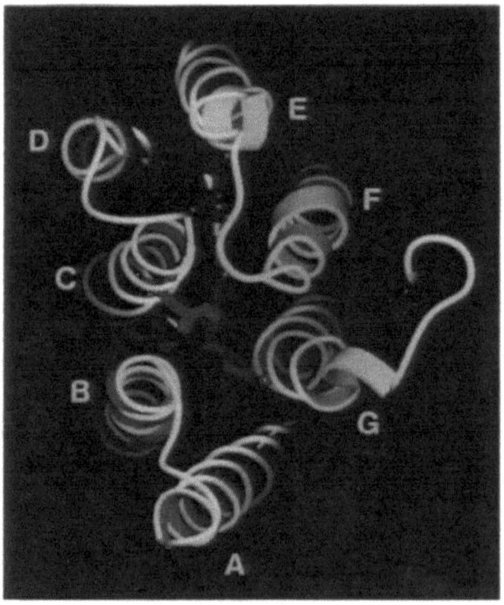

Fig. 5.3. View at the cytoplasmic surface onto bacteriorhodopsin. Helices E-G and A have stronger tilt angles. Retinal (purple) is nestling in a membrane pocket. D96 and D85 are marked in red in helix C.

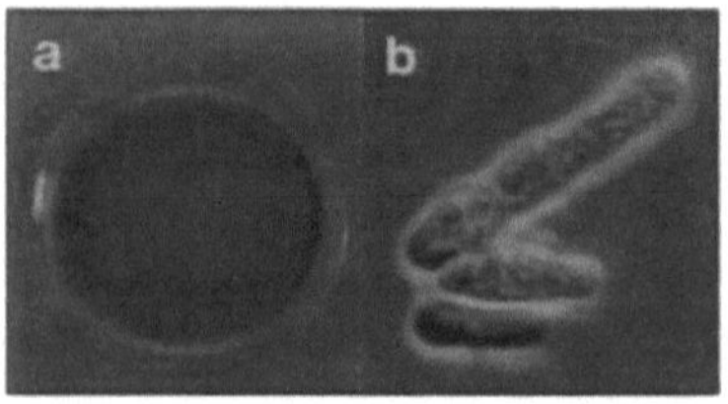

Fig. 5.5.
a. Wet cell pellet of fission yeast cells expressing bacteriorhodopsin.
b. *S. pombe* cells grown in EMM without thiamine. Two cells which overexpress bacteriorhodopsin have proliferated intracellular membranes and show an increase in size. One cell has a normal shape.

Index

MIX
Papier aus verantwortungsvollen Quellen
Paper from responsible sources
FSC® C105338

If you have any concerns about our products,
you can contact us on
ProductSafety@springernature.com

In case Publisher is established outside the EU,
the EU authorized representative is:
Springer Nature Customer Service Center GmbH
Europaplatz 3, 69115 Heidelberg, Germany

Printed by Libri Plureos GmbH
in Hamburg, Germany